Female plants of the alga *Chara corallina* (formerly *C. australis* var. *nobilis*):
cultured material. The scale mark represents 10 mm. (Photo: B. T. Lester.)

Frontispiece

THE PHYSIOLOGY OF GIANT ALGAL CELLS

A. B. HOPE

Professor of Biology, Flinders University of South Australia

N. A. WALKER

Associate Professor of Biology, University of Sydney

CAMBRIDGE UNIVERSITY PRESS

CAMBRIDGE UNIVERSITY PRESS
Cambridge, New York, Melbourne, Madrid, Cape Town,
Singapore, São Paulo, Delhi, Tokyo, Mexico City

Cambridge University Press
The Edinburgh Building, Cambridge CB2 8RU, UK

Published in the United States of America by Cambridge University Press, New York

www.cambridge.org
Information on this title: www.cambridge.org/9780521279314

First published 1975
First paperback edition 2011

A catalogue record for this publication is available from the British Library

Library of Congress Catalogue Card Number: 74–77832

ISBN 978-0-521-20513-9 Hardback
ISBN 978-0-521-27931-4 Paperback

Contents

v

vi

Preface

Giant algal cells have provided experimental material for the physiologist for a little more than half a century: the interaction of physics with biology is several hundred years old. Both physicists, we have written of those parts of the physiology of giant algal cells that interested us, omitting many things of interest to biochemists or geneticists. In this chosen, restricted field, we have kept in mind that we were summarising the latest work, for our colleagues, and introducing the field to others. At times the conflict between these imaginary readers will be apparent to the real one.

When this book was mooted, the field of biophysical physiology of giant algal cells was full of active workers again, and full of unresolved problems. We thought, as time passed, that the book would be able to report more and more of them to have been successfully solved. Nine years later, many an old dilemma stands, less threatening than before, its familiar horns worn by the passing summers and winters. New problems claim attention: as we write, an interesting clash of paradigms is in progress. For many years, the models in use in plant electrophysiology came from animal physiology; now that Mitchell has turned so much of biochemistry into a branch of membrane biophysics, the models are beginning to be Mitchellian. Here, as elsewhere, we can only urge the reader who wants clear-cut decisions to wait-and-see.

We thank our colleagues who helped us during the preparation of the book: especially G. P. Findlay, who read the manuscript, and whose criticism was of great use to us; and also those who without reading the manuscript delicately hinted that it had better be good. Many gave us access to their unpublished work, for which we are grateful – C. E. Barr, P. H. Barry, H. G. L. Coster, W. J. Cram, G. P. Findlay, G. E. Fogg, P. B. Green, A. W. D. Larkum, R. McC. Lilley, L. J. Mullins, J. D. Pickett-Heaps, M. G. Pitman, F. A. Smith, W. J. Vredenberg, E. J. Williams, and U. Zimmermann.

A*

For the photographs of algae that grace the volume we thank B. T. Lester and J. P. Fairburn of Sydney University; for the electron micrographs we thank T. E. Bostrom, A. W. D. Larkum, J. D. Pickett-Heaps and M. Vesk. For many of the line drawings, our thanks to D. J. Stanley of Sydney University.

Finally we are happy to acknowledge the following for permission to reproduce published journal figures:

Botanical Society of Japan [*Botanical Magazine, Tokyo*: Fig. 12.2 (*a*)].

CSIRO [*The Australian Journal of Biological Sciences*: Figs 5.9, 7.4 (*a*), (*b*), 7.5 (*c*), 7.6, 7.8, 8.3].

Elsevier Publishing Company [*Biochimica et biophysica Acta*: Figs 6.8 (*a*), (*c*), 10.5, 10.6].

Japanese Society of Plant Physiologists [*Plant and Cell Physiology*: Fig. 12.5 (*b*)].

Macmillan Journals Ltd [*Nature, Lond.*: Fig. 2.4].

New Phytologist [Figs 10.2, 10.3].

Oxford University Press [*Journal of experimental Botany*: Figs 6.4 (*b*), 11.1].

Physiologia Plantarum [Fig. 4.1].

Rockefeller University Press [*Journal of general Physiology*: Figs 2.1, 2.5, 6.4 (*a*), 7.4 (*c*), 12.5 (*a*)].

Springer-Verlag [*Protoplasma, Handbuch der Pflantzenphysiologie*: Figs 2.2, 2.3, 3.1, 3.4, 12.2 (*b*), 12.3, 12.4].

A.B.H.
N.A.W.

November 1973

Symbols used

A Area (m^2)

a Semi-major axis of ellipse (m)

b Semi-minor axis of ellipse (m)

c_j^o Concentration of species j in phase o (M)

D Diffusion coefficient $(m^2\,s^{-1})$

F Faraday's constant $(F = 9.65 \times 10^4\,C\,mol^{-1})$

G Gibbs free energy (J)

G_M Specific chord conductance of membrane $[G_M = \varDelta \mathcal{J}_M / \varDelta \psi_M]$ $(S\,m^{-2})$

g_M Specific differential conductance of membrane $[g_M = \partial \mathcal{J}_M / \partial \psi_M]$ $(S\,m^{-2})$

g_j Specific differential conductance of membrane due to movement of species j $(S\,m^{-2})$

g Acceleration due to gravity $(g = 9.81\,m\,s^{-2})$

H Enthalpy (J)

$\mathcal{J}$ Current density $(A\,m^{-2})$

$\mathcal{J}_v$ Volume flow per unit area $(m\,s^{-1})$

K Partition Coefficient of substance between oil and water, or membrane and water

k Rate constant (s^{-1})

$\boldsymbol{k}$ Boltzman's constant $(\boldsymbol{k} = 1.381 \times 10^{-23}\,J\,K^{-1})$

L_p Hydraulic conductivity $(m\,s^{-1}\,Pa^{-1})$

l Length (m)

M Relative molecular mass [Molecular weight]

$\mathcal{N}$ Avogadro's number $(\mathcal{N} = 6.022 \times 10^{23}\,mol^{-1})$

$\boldsymbol{P}$ Hydrostatic pressure (Pa)

P_K Permeability of membrane to species K $(m\,s^{-1})$

P_{os} Osmotic water permeability of membrane $(m\,s^{-1})$

P_d Diffusive water permeability of membrane $(m\,s^{-1})$

pK_a Negative logarithm (base 10) of acid dissociation constant

Q_a Quantity of given species in phase a, per unit area of cell surface $(mol\,m^{-2})$

R Gas constant $(R = 8.31\,J\,mol^{-1}\,K^{-1})$

r Radius (m)

r_{jk}	Resistance coefficient in force-flow equation
S_a	Specific radioactivity of a given species in phase a (s^{-1} mol^{-1})
T	Absolute temperature (K)
t	Time (s)
V	Volume (m^3)
$\overline{V}$	Partial molar volume (m^3 mol^{-1})
v	Mean velocity (m s^{-1})
Υ_a	Quantity of radioactivity in phase a, per unit area of cell surface (s^{-1} m^{-2})
α	Ratio of sodium permeability to potassium permeability ($\alpha = P_{Na}/P_K$)
γ	Ratio of chloride permeability to potassium permeability
Δ	Change in (e.g. ΔG etc.)
μ	Chemical potential (J mol^{-1})
$\bar{\mu}$	Electrochemical potential (J mol^{-1})
Π_i	Osmotic potential in phase i (Pa)
σ	Reflection coefficient of membrane for a given species
ϕ_{jab}	Unidirectional flux of species j from phase a to phase b (mol m^{-2} s^{-1})
ϕ_K	Flux of species K across membrane (mol m^{-2} s^{-1})
Ψ_w	Water potential (Pa)
ψ_{io}	Electrical potential of phase i with reference to phase o (V)
[ ]$_i$	Concentration in phase i [e.g. $[K^+]_i$]
ψ_K	Nernst equilibrium potential for species K (V)
z_j	Algebraic valency of ions j

Abbreviations

ADP	Adenosine diphosphate
AHW	Artificial Hickling water
AMP	Adenosine monophosphate
AP	Action potential
APW	Artificial pond water; for composition see table 5.1
ASW	Artificial sea water
ATP	Adenosine triphosphate
CCCP	Carbonyl cyanide m-chlorophenyl hydrazone
DCCD	Dicyclohexylcarbodiimide
DCMU	3'-(3,4-dichlorophenyl), 1', 1'-dimethylurea
DCPIP	2,6-dichloro-phenol indo-phenol
DNP	2,4-dinitrophenol
DSPD	Disalicylidenepropanediamine
FCCP	Carbonyl cyanide p-trifluoromethoxyphenyl-hydrazone
FP	Flavoprotein
FPW	Flinders pond water; for composition see table 5.1
HOQNO	4(n-Leptyl)-hydroxyquinoline-N-oxide
IAA	Indole-acetic acid
$NADH_2$	Nicotinamide adenine dinucleotide (reduced)
$NADPH_2$	Nicotinamide adenine dinucleotide phosphate (reduced)
NEM	N-ethyl-maleimide
OP	Osmotic potential
p.d.	Electrical potential difference
PCMB	Parachloro-mercuribenzoate
PCMBS	Parachloro-mercuribenzoate (sulphonic acid)
PMS	Phenazine methosulphate
PSI	Photosystem I
PSII	Photosystem II
Q_{10}	Ratio of rates at temperatures differing by 10 K
SCC	Short circuit current
SW	Sea water
TP	Turgor pressure

CHAPTER 1

Giant cells as experimental material

Algae with giant cells

Physiologists have often profited from the survival of animals and plants which have cells of more than ordinary size. It is difficult to imagine that the physiology of nerve and muscle could have reached its present state without the existence of the giant axon of the cephalopod. Though rapid conduction of electric stimuli does not appear to have necessitated large cells in plants, many algae have derived the simple advantage of large thallus size from their possession of giant cells. Long before Young (1936) drew attention to the squid giant axon, physiologists had found *Valonia* and *Nitella*, and had begun work on many of the problems discussed in this book.

In some algae the evolutionary sequence is open to doubt, and it may not be certain whether the giant cells arose by the simple increase in size of one cell or by the suppression of cross-wall formation in a tissue. Steward & Sutcliffe (1959) raised this question, which they stated in the form of the problematical homology of the vacuole of the normal plant cell and that of the coenocyte. These days the question has been allowed to lapse: the common assumption is that cell and coenocyte are homologous and that their vacuoles are too. Here we will simply call coenocytes 'cells' for brevity.

Taxonomy

A taxonomic chart of most genera exhibiting giant cells is given in fig. 1.1.

The Charophyta

Six related genera form this group, which has occupied an uncomfortably variable taxonomic niche. Indeed it is a matter of semantic convenience, rather than a statement of fact or of well-grounded faith, when we include them among the algae.

They have been persuasively classified as an order, Charales, of the algal class Chlorophyceae (Fritsch, 1965), as an independent division, Charophyta, by the editors of *Biological Abstracts*, and as a division related to the bryophytes rather than to the algae. Rather than dispute a matter of taste, we accept them as a full division of the algae, the Charophyta, following Round (1971).

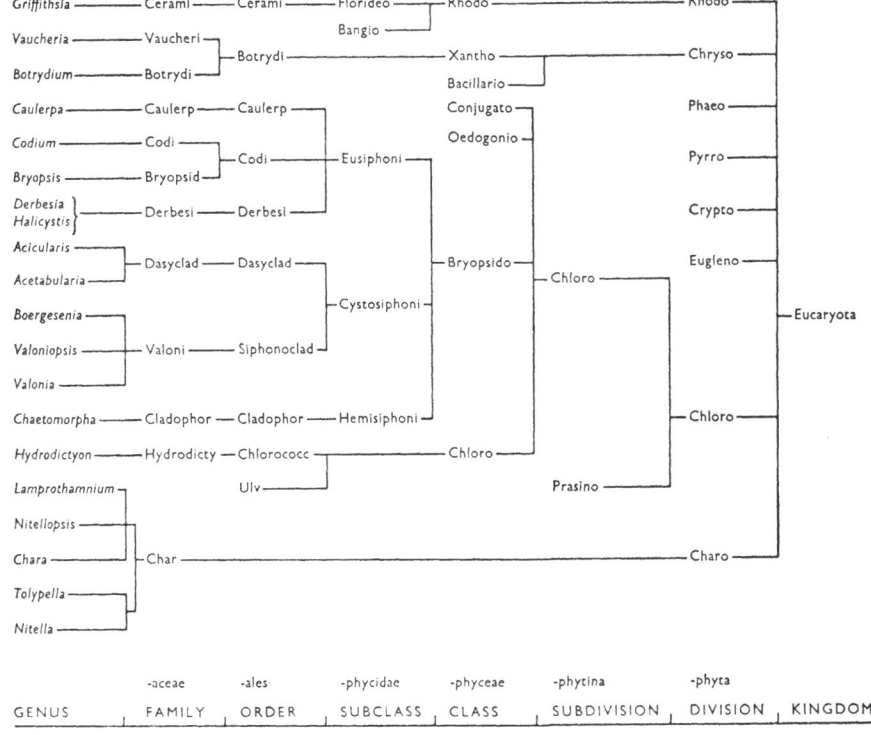

Fig. 1.1   A taxonomic chart of many of the giant-celled algae that have been used in experiments.

Undoubted charophyte oospores have been found as far back as the Ordovician $(-4 \times 10^8$ years): this tribute to their durability has not clarified the taxonomic relationships of the division.

A complete conspectus of the charophyte species used in research is given by Wood (1952); details and drawings of all species are to be found in Wood & Imahori (1965). These authors, using normal criteria, have lumped together many once-different species, notably *C. australis* and *C. corallina*; on the other hand, the breeding experiments of Proctor (1971) suggest that on this criterion the many charophyte species should be still further split. The group consists of water plants, basically freshwater, though some species of *Chara* and *Lamprothamnion* can be found in brackish and salt water. The group has a world-wide distribution in all but arctic zones. In most places where there are bodies of freshwater, running or stagnant, permanent or seasonal, they can be found. In many places they are common plants, forming clumps or meadows at depths to several metres on sandy or muddy bottoms where their rhizoids anchor them.

The visible plant is some centimetres to several metres high, green in colour, composed of cylindrical, unicellular internodes and small multicellular nodes (fig. 1.2). This visible plant is haploid; some species are dioecious and others monoecious. Ova and motile antherozoids are produced in decorative and intricate sex organs (plate 1); fertilisation gives rise to a dormant oospore, resistant to drying, and encased in a shell of calcium carbonate. This zygote alone is diploid; meiosis occurs on germination, and further cell division produces a protonema and a rhizoid-initial. The mature plant begins as a lateral on the protonema.

The usefulness of these plants in the laboratory has depended upon the ability of the internodal cells to survive isolation from the plant, and on the large size and regular shape of these cells. In *C. corallina*, the varieties used in research have axial internodes about 1 mm in diameter and up to 15 cm long when grown in laboratory culture; collected from the field they usually have larger diameters than this. These internodes, and the 'leaf' internodes, approximate to cylinders in shape, and in the ecorticate species they provide seemingly ideal material for many physiological observations and manipulations. While the use of a corticated species is not quite unknown (e.g. Gaffey & Mullins, 1958) most workers have felt that the ecorticate forms provide a large enough number of problems without gratuitous geometrical complexity. As isolated, the internodes of the ecorticate kinds have a disk of nodal cells at each end. For measurements of ionic fluxes, many of the nodal cells may be

3

carefully dissected away, but usually simpler precautions are taken to avoid spurious results from their presence.

The relationship of the single cell to the whole plant is a question taken up again in the chapter on protoplasmic streaming: here it may be useful to mention some properties of the plant which are suggestive.

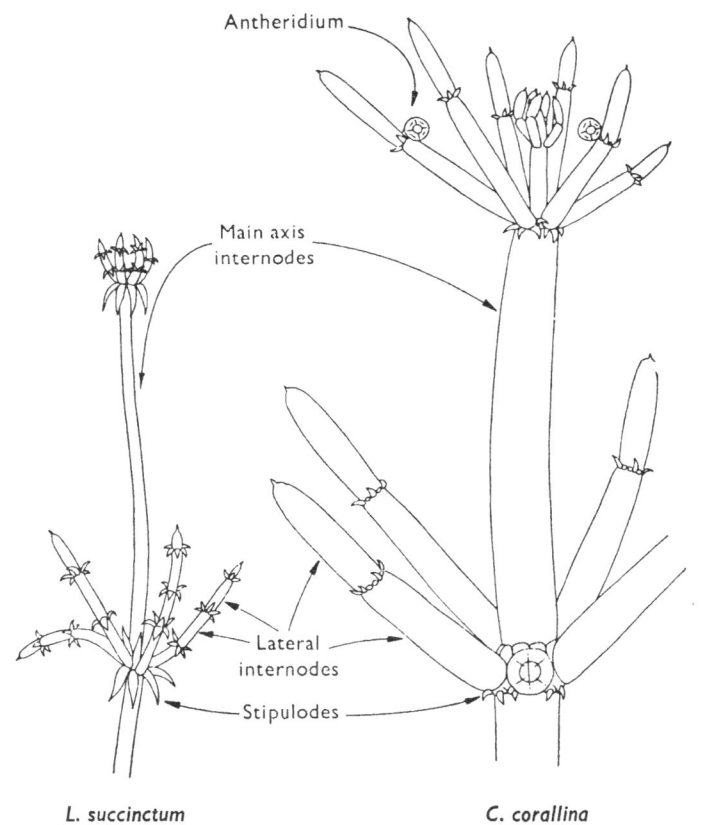

L. succinctum   C. corallina

Fig. 1.2   Sketches of growth habit of plants of the algae *C. corallina* and *Lamprothamnium succinctum.*

The nodal cells are uninucleate and the internodes multi-nucleate, having become so by 'nuclear fragmentation' rather than by repeated mitosis. This seems to have been relatively little studied, but it has no obvious implications for physiological work.

4

The plant exhibits growth correlations of the sort normally associated with higher plants. The main axis grows at its apex, and this apex inhibits the development of secondary apices which arise at each node, since the removal of the apex from the main axis results in the development of one or more of these secondary apices. Rhizoid development is promoted by the apex, and it occurs at the lowest node of the isolated shoot.

The Bryopsidophyceae

These are mostly marine algae, the order being characterised by a coenocytic habit which certainly seems to be the result of failure to form cross-walls. From simple spherical shapes such as those of *Protosiphon* and *Halicystis* (the former never used in physiology, the latter quite popular) the order includes as well the ambiguous genus *Valonia*, whose shape may be relatively simple (but irregular) or quite complicated. *Valonia* represents an approach to multicellularity, though it has been used only as a unicellular preparation. Some genera have trabeculae on the intercellular walls. Further complexity is offered by such genera as *Caulerpa* and *Bryopsis*, which have shapes of some regularity and beauty. They, and the more complicated *Dasycladus*, are objects to marvel at rather than to compute the surface area of. The easy ones are given in table 1.1. *Acetabularia* and *Acicularia*, simplest of the Dasycladaceae, are famous uni-nucleate preparations for work on nuclear–cytoplasmic interaction, and there are several studies of the ionic relations of *Acetabularia*.

TABLE 1.1 *Formulae for surface area*

Chara, Nitella	Cylinder	$\pi d(l+d/2)$
Hydrodictyon, Valonia	Sphere	πd^2
Chaetomorpha	Truncated sphere	$\pi dl + \pi \delta^2/2$
Griffithsia pulvinata	Prolate spheroid	$2\pi[b^2+(ab \cdot \arcsin \epsilon)/\epsilon]$

Where d = cell diameter, l = cell length, δ = diameter of end of truncated sphere, a and b = semi-major and semi-minor axes and $\epsilon = (1-b^2/a^2)^{1/2}$.

Of these plants, *Valonia* and *Halicystis* have been most used for physiological work. Indeed, *Valonia* may be said to have begun electrophysiology, since its cell sap was analysed before the turn of the century, and its vacuole potential was measured before that of any other cell. *Halicystis* attracted the attention of Blinks (1929), who studied the vacuole potential and ion concentrations. Both genera have large cells with a thin layer of gel cytoplasm containing numerous nuclei and chloroplasts. *Valonia* reproduces by liberating zoospores, each of which regenerates a new thallus like that of the parent. *Halicystis* liberates gametes and the resulting zygote forms a plant of *Derbesia*, which in due course liberates zoospores destined to be *Halicystis* plants. Genera related to *Valonia* are *Valoniopsis* and *Boërgesenia*; a little work has been done on each.

The Chlorococcales

The Chlorococcales have a single genus with the giant-celled habit, *Hydrodictyon*. The spheroidal to cylindrical coenocytes form a network containing as many as 256 or 512 individuals per net. Each of these individual cells can form internal zoospores, which aggregate inside the old cell wall to form a new net. Again, the cell has a thin layer of solid, chloroplast-packed cytoplasm and a large central vacuole. In *H. africanum* the net is easily separable into its individual cells.

The Ceramiales (Rhodophyta)

So far the only red algal genus much used for its giant cells has been *Griffithsia* of the family Ceramiales. The advantages of using a cell whose photosynthetic apparatus has two light reactions fed by widely different pigments are only now beginning to be realised. *Griffithsia* has a variable, bushy habit; the multi-nucleate cells generally are some 2 to 4 mm long and perhaps half as wide, sometimes approximating in shape to a prolate spheroid (plate 2). Like all other giant plant cells, they are vacuolate, with a thin layer of cytoplasm lining the wall. The plastids are numerous. The plants are collected at low tide, being among the many red algae that do not grow at the great depths at which they are supposed to have an adaptive advantage.

Ultrastructural features

The plants considered here all show the general ultrastructural features expected of eucaryotes, with essentially normal organelles, and with some range of chloroplast structure (Bisalputra, 1974).

There has not been a very large number of purely ultrastructural studies on these cells, and those that have been undertaken as an adjunct to physiological work have sometimes been perfunctory. Modern methods of fixation – notably the use of glutaraldehyde – seem to preserve algal structure reasonably well, and the time is ripe for a considerable increase in our knowledge of giant cell ultrastructure. Good beginnings have already been made with *Chara* (Pickett-Heaps, 1967) and with *Hydrodictyon* (Marchant & Pickett-Heaps, 1970, 1972).

Of the few features which should be noticed here the most important is the arrangement of phases, that is, the arrangement of cytoplasmic membranes inside the cell. Here an interesting contrast emerges between charophyte cells, with their layer of cytoplasm about 10 μm thick, and marine cells like *Griffithsia* where the bulk of the cytoplasmic volume is occupied by chloroplasts (plate 3), and the 'ground' plasm is a thin layer or a spongy mass deeply penetrated by vacuolar contents. In neither case does the picture give much hope for the insertion of microelectrodes into the cytoplasmic phase. *Valonia*, *Valoniopsis* and *Chaetomorpha* are also difficult subjects for the insertion of microelectrodes into the cytoplasm, though none has such a very thin cytoplasmic layer.

In charophytes, kinetic studies have made it important to know the arrangement of phases within the cytoplasm. In the most recent investigation, Costerton & MacRobbie (1970) showed, like earlier workers, no sharp dividing line between the peripheral gel cytoplasm which contains chloroplasts, mitochondria, dictyosomes and endoplasmic reticulum (e.p.r.), and the internal, mobile cytoplasm. Indeed, as had others, they found e.p.r. in the stationary gel connected with the e.p.r. in the mobile sol. In the gel the e.p.r. had the form of flat plates, while in the solution the form was that of interconnected tubules. Their pictures show profiles of the e.p.r. swollen to form small vacuoles (cf. plate 5*a*), and they suggest, very plausibly, that the e.p.r. may be an important route for salt transport to the vacuole. It certainly seems to be the route for the transfer to the vacuole of the large protein crystals which occur there frequently in charophyte cells. It is not however clear that the e.p.r. can transfer its contents directly to the vacuole by fusing with the tonoplast, since the protein crystals are often, in the vacuole, surrounded by membrane or remnants of membrane.

7

Another charophyte curiosity was announced briefly by Barton (1965a, b) and by Crawley (1965). In young 'leaf' internodal cells of *Chara* and *Nitella*, the plasmalemma contains oval organelles (plate 4) with an organisation strongly resembling that of the chloroplast prolamellar body (Gunning, 1965). Its interest here is chiefly that it provides a local increase in membrane area of the order of tenfold, and that it may occur over large fractions of the total membrane area. A function in cell wall deposition is possible, though an autoradiographic study with labelled glucose (Barton, 1968) makes this rather unlikely. An ion transport role is possible, but there is no normal ion transport for which an increase in membrane area would offer an advantage. No association has been found between the organelle and the acid or alkali excreting zones of the cell (Vesk & Walker, unpublished). An analogy with the prolamellar body has suggested to R. Barton (personal communication) that the organelle may represent a store of membrane material that will be converted into normal plasmalemma as the cell rapidly elongates. The apparent association between the organelle and mitochondria would then reflect the role of the latter in lipid metabolism rather than in energy supply.

Most of the giant cells mentioned in this chapter appear to live in physiological isolation, although in *Valonia* the existence of trabeculae on the intercellular walls suggests intercellular transport. In charophytes the giant cells are clearly part of an organised and differentiated plant, and extensive plasmodesmata (plates 5a,b) are found to join the cytoplasm of neighbouring cells (Spanswick & Costerton, 1967). They have average external diameters ranging from 62 to 71 nm, and comprise 0.06 to 0.15 of the total area of the intercellular wall, the larger fractions occurring in immature cells and in walls separating nodal cells from each other. In charophytes the plasmodesmata do not seem to be associated with elements of the endoplasmic reticulum. Those in node–node walls are relatively simple, but those in the thicker node–internode walls have a complex geometry, showing simple anastomoses or joining to form complex sinuses (plate 5b). The material in these sinuses is bounded, like that in the plasmodesmata, by an extension of the plasmalemma. It is uniformly stained by current electron microscopical procedures.

8

This section can unfortunately do little but draw attention to the unsatisfactory state of its subject matter. As would be expected, most of the pioneering physiology and electro-physiology was done on collected material, and hence work on marine algae was concentrated on such places as Naples, Woods Hole, Bermuda and La Jolla. Workers at other places have sometimes arranged regular collections of marine plants to be shipped. Freshwater plants too have frequently been collected from the field by the experimenter. Such collecting, enjoyable though it is to the laboratory worker, is a quite unsatisfactory way of obtaining material for experiment. There is a marked seasonal variation in collected plants of the Charophyta; in particular *C. corallina* shows a high mortality rate in the laboratory when gathered in summer. In some seasons the target plant is simply not to be found, either because its habitat has dried up, or because there has been unexpected pollution of the collecting grounds.

For these and other obvious reasons the culture of the experimental material under constant conditions must be the goal of the serious worker. It is far from easy to achieve this in many cases.

Charophyta

Although the spores can be surface-sterilised and will germinate, these plants have usually been grown from 'cuttings' of one or more internodes. It is difficult to grow them in a chemically defined medium, and a common finding is that if one tries to do this, the cutting will only succeed in growing when it comes from a plant grown in soil; cuttings from plants grown in the defined medium will not develop into plants.

For most experimental work the culture method is to plant the cuttings in soil or mud at the bottom of a large jar or aquarium tank, and to fill with tap water or some solution of desirable composition. The writers use river mud from locations where *Chara* flourishes, filling the culture tank with solution of the following composition: (mM) 0.5 $NaHCO_3$, 1.0 NaCl, 0.2 $NaNO_3$, 0.017 KH_2PO_4, 0.05 K_2SO_4, 0.1 $MgSO_4$, 0.1 $CaCl_2$ and (μM) 3.6 $FeSO_4$, 0.91 $MnSO_4$, 0.76 $ZnSO_4$, 0.014 $(NH_4)_6$

Mo_7O_{24}. This is a modification by Hope of a recipe given by Barr & Broyer (1964).

Plants grow vigorously in this medium, being kept at about 20 °C and in various light regimes. *Nitella translucens* can be grown in the same way, and – in a medium of more salty composition – so can *Nitellopsis obtusa*. Since the plants are grown from cuttings, one may raise a clone from a single plant. The hope that this would produce plants that are more physiologically uniform than usual does not seem to have been borne out.

With *Nitella flexilis* P. B. Green (personal communication) has used what seems to be a more successful method. Selected soil and distilled water are autoclaved together in one-litre glass cylinders, and a cutting established in each. Two weeks later the vigorously growing plant is harvested, and some of it used for further cultures. The method depends critically on a suitable supply of soil for the species to be cultivated, but when working well it produces a regular supply of uniformly growing material. Forsberg (1965) has grown the corticated *Chara zeylandica* in chemically defined culture, but the method has not been widely adopted; Barr & Broyer (1964) have similarly grown *N. clavata* but Barr does not now routinely use the method, preferring to add soil decoction to his cultures (Rent, Johnson & Barr, 1972).

Green (1968) alternated two weeks growth on soil water medium with two weeks on Forsberg medium, producing plants free of epiphytes.

Bryopsidophyceae

These plants have been less widely cultured in the laboratory, but physiologists interested in nuclear–cytoplasmic interactions have grown *Acetabularia* in the laboratory for some time (Keck, 1964). It is grown in a medium of natural sea water, plus soil extract and some additional salts.

Gutknecht (1966) has used this method for *Valonia*, and has raised cells to several millimeters in diameter in the laboratory; to obtain cells several centimeters in diameter, collection has been the only method available.

Chlorococcales

Hydrodictyon has been used mostly by workers who have cultured it; Raven's method (1967) is the same as that used by earlier workers at Cambridge. The orange zygotes are germinated in a soil–water medium prepared by twice steaming in Kilner jars and then storing for one month (Northcote, Goulding & Horne, 1960). The best growth of the coenobia is achieved at about 13 °C.

Cell operations

The simplest operation that can be carried out on giant cells is the removal of sap for analysis. Large spheroidal cells are the simplest in this respect, the gel-like cytoplasm being unlikely to contaminate the sample. The extent of this contamination is not often assessed, however. *Valoniopsis* sometimes offers problems to the sap-sampler, since its vacuolar contents may be mucilaginous, difficult to express and to take up into a capillary micropipette. The sap of *Caulerpa* is also mucilaginous.

Valonia offers the useful possibility of instant wall preparation, exploited by Gutknecht (1966). If seawater is admitted to the vacuole the protoplast converts itself over the course of an hour into tiny rounded aplanospores, leaving most of the area of the cell wall clear. This allows rapid comparison of the properties of the live cell and of the isolated wall (chapter 3).

With the Charophyta there are other problems and possibilities. The cell wall may be scraped clean of protoplast by cutting off the ends of the cell and running a blunt straightedge along it. These wall preparations may be reinflated to cylindrical shape or cut open to form ribbons.

If the end is cut off a turgid cell, some of the contents are ejected by the internal hydrostatic pressure. Use of a really sharp blade helps here, but the important trick is to blot the cell dry, and to blow on it for a moment to reduce its turgor. In this way, the cell may be cut open without its collapsing. Two methods may now be used in the attempt to separate sap from cytoplasm. MacRobbie's method (1962) is to cut an end from the cell, to thread it onto the tapered end of a Pasteur pipette, to cut the other end and slowly to blow an air bubble through the cell. This expels the sap and leaves the cytoplasmic layer largely intact, though it is not clear that the plane of separation coincides exactly with the tonoplast. A preferred method of obtaining a sap sample is to cut off one end (of a reasonably long cell) and hold it vertically by the other.

A drop of a few microlitres will run out; very gentle pressure with fingers or forceps may enlarge the size of the drop, but there is then the probability of adding some of the flowing cytoplasm to it. In using this method the cytoplasmic content may be calculated from an analysis of the sap sample and of the remainder of the cell – both must be weighed or have their volume measured.

If the whole cell is placed in a centrifuge and gently spun, for example at 200 g for 10 min, the flowing cytoplasm may be collected at the centrifugal end of the cell without dislodging the layer of chloroplasts. In this way a microlitre or so of clear cytoplasm can be got for analysis. It is obvious that this sample may be contaminated by sap vacuoles. This has been suggested by Kishimoto & Tazawa (1965a), and there is some evidence for it in their work. Low power freeze etching studies with an electron microscope would help to resolve this.

If the halves of a *Chara* cell are bathed in solutions of different osmotic potential, the water flow which is carried through the cell vacuole will sweep solute molecules out of the inflow end and up to the outflow end (transcellular osmosis, chapter 3). Kamiya & Kuroda (1956a) learned to tie off the two halves of the cell with a silk ligature – the result is two cells whose vacuolar contents have different osmotic potentials.

Further, Tazawa (1964) was able to perfuse the vacuole by the following delicate procedure: a cell whose turgor has been reduced to zero, and whose ends were bathed in isotonic solutions, had these ends cut off. An extremely small pressure head between the ends drove a slow flow of solution through the cell. If this solution contained neutral red or phenol red its progress through the cell could be followed. When the sap had been completely replaced the protoplast sleeve was available for measurements, or it could be tied off at each end to form a new cell with an artificial vacuolar sap.

Strunk (1970) found with *N. clavata* that he could attach the cut ends of the cell to glass cannulae by allowing the cell wall to dry onto the glass. Although it is not clear how long such cells will survive, this technique seems worthy of further development. Spyropoulos (1972) describes a more elaborate apparatus for perfusing *Nitella*.

Cellular perfusion had long been possible with some marine algae. *Valonia* was perfused by Damon (1929) and this technique was used on *Halicystis* by Blinks (1929). Into these large coenocytes one may insert two large capillaries, and a slow flow of artificial sap may be kept up for long periods.

Membrane perfusion seems not to have been achieved yet with any of these cells as it has been with the giant axon, perhaps because it has not been tried with enough persistence.

A number of pretty techniques has been evolved by investigators of protoplasmic streaming. Chloroplasts have been dislodged by jarring the cell reproducibly in a cavity in a small hammer (Kamiya & Kuroda, 1964). A better technique (R. D. Allen, personal communication) is to irradiate the cell with a focussed, small beam of ultraviolet light. This causes the chloroplasts to leave the irradiated area, which becomes a window into the cell.

If the end is cut off a cell under the surface of an isotonic solution, streaming may continue, and a drop of clear, flowing cytoplasm will depend from the cell and then fall (Kamiya & Kuroda, 1957; 1965). This elegant technique for isolating undisturbed cytoplasm has been recently applied to studies of the electrical properties of the surrounding membrane (Inoue, Ueda & Kobatake, 1973).

The hydrostatic pressure inside a growing *Nitella* cell has been monitored in a most elegant set of experiments by Green (1968). A micro-capillary inserted into the cell vacuole leads to a closed capillary containing a bubble whose volume measures the hydrostatic pressure.

Electrical measurements

The electrical properties of the cell membranes are accessible to techniques using external contacts, for which giant cells are necessary, and to techniques using inserted electrodes, usually salt-bridges, for which giant cells are most useful.

Measurements of electric potential differences in cells have generally been interpreted in terms of membrane properties. Thermodynamicists stress that a potential difference such as that across a membrane can only be inferred after an act of faith about the cancellation of the liquid-junction potentials in the salt-bridges (Spanner, 1964). Faith is necessary because liquid junction potentials may be calculated, but not independently measured. For the case of a salt-bridge in a plant cell vacuole, such a calculation is likely to be reliable to within a few millivolts; when the bridge is in cytoplasm, one can proceed with less confidence, since cytoplasm contains charged polymers as well as crystallites. In such a phase the concept of an electric potential characteristic of the whole phase becomes as fuzzy as the calculations.

To these problems, which face all measurements of liquid-phase potentials, plant cells add an extra difficulty which arises from the positive hydrostatic pressure of their contents. When an open-tipped micro-salt-bridge is inserted into such a cell, the contents will gush up into the electrode unless and until the tip is blocked by some more or less solid body. This poses a dilemma: if the tip is blocked, the liquid junction cannot be calculated with any hope of precision, since the tip is equivalent to a membrane of unknown properties; if the tip is not blocked, the turgor of the cell is inevitably reduced to zero. One might expect the unknown tip potentials to be smallest when the electrolyte filling the salt-bridge most nearly resembles the cell contents: potential differences so measured are not much different from those measured with the usual 3 M KCl.

Gushing can be made use of in practice when one desires to measure the potential of the vacuole in charophyte cells. A microelectrode of tip diameter 5–20 μm is preferred, the momentary gush enabling it to break through the cytoplasm into the vacuole, where small crystals, etc. plug the tip quickly. A microelectrode of diameter 1 μm or less will almost always have its tip in the cytoplasm, carrying a sheath of cytoplasm with it as it is pushed into the cell. Marine coenocytes with gel-like cytoplasm generally allow a 2 μm electrode to enter the vacuole easily – a larger one may allow prolonged gushing. Contrariwise it may be extremely difficult to get a micro-electrode tip into the cytoplasmic layer at all.

Apart from junction potential artifacts, there is the possibility of damage to or change in cells caused by the insertion of electrodes. Such a change might most obviously be brought about by a loss of turgor pressure, by an electrical leak where the electrode has burst through a membrane, or by the diffusion of salt out of the electrode tip. The first is small in giant cells, provided that the tip becomes blocked: Green & Stanton (1967) have shown that even the insertion of their relatively large manometer tubes into *Nitella* cells produced a loss of turgor much less than 20 kPa. Checks that there is not much electrical leakage around an inserted probe were made by Hope & Walker (1961) and Skierczynska (1968) for charophyte cells. Though there have been reports of cells needing 12 hours in which to reach a constant, high resistance (Skierczynska, 1968; Spanswick, 1970b) our experience has been that one to two hours suffices with *Chara* and *Griffithsia*. In charophyte cells, a seal of unknown composition is synthesised around inserted electrodes (Umrath, 1932; Walker, 1955) which will ultimately cover the glass tip and exclude it electrically from the 'inside' of the cell. This seal presumably consists of new cell wall and membrane. Microelectrodes can often be withdrawn at this stage without stopping the protoplasmic streaming and without escape of cell contents. Other indicators of normal behaviour of cells with implanted microprobes are (a) normal cytoplasmic streaming (charophytes) and (b) the same influx and efflux of ions as intact cells. Microscopic control of the whereabouts of the tip is sometimes possible in the more transparent coenocytes. Electrical criteria for the location of the tip can be worked out – see the work of Findlay and of Spanswick – and are obviously necessary to careful work.

Electrical techniques

The equipment needed to measure and to record the resting potential and resistance of a large plant cell is briefly:
1. Cell holder with irrigation.
2. Microelectrodes (microcapillary salt-bridges); micro-electrode puller.
3. Calomel half-cells.
4. Micromanipulators.
5. Dissecting or binocular microscope with low power as well as high power objectives.
6. High input-impedance (~ 10 GΩ), differential voltage amplifier.
6'. Voltage amplifier ('operational amplifier').
7. Pen recorder, 1–2 channels.
8. Pulse or direct current generator.
9. Potentiometer for backing-off, and switches.

Although the home-made thermionic valve electrometer played its part in the fifties as it had done a generation earlier the measurements were made easier by the production of good commercial electrometers. The high input-resistance, differential-input, model 603 electrometer of the Keithley Instrument Co. was outstanding in its field, superseded only recently by solid-state amplifiers.

The general arrangement of 1–9 is shown in fig. 1.3. With this set-up, the resting potential of the vacuole can be recorded until the microelectrode has become sealed off from the 'inside' of the cell (see above). At any time, the p.d. can be backed off by 9, the recorder sensitivity increased and the change in p.d. recorded on switching on a small current from 8. The current may be registered as a change in p.d. across the 10 kΩ resistor. The readings lead to values of the 'membrane conductance' (cf. Walker, 1960).

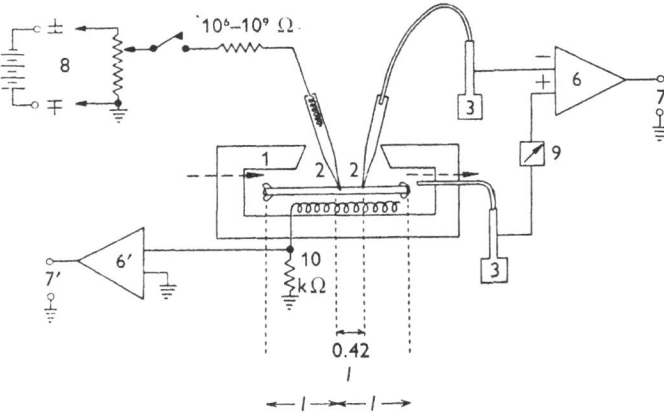

Fig. 1.3 Set-up for measuring cellular potentials: a charophyte cell is shown. The components are explained in the text. *l* is the half-length of the cell.

The recorder will need a suitably fast response (say 5–10 Hz) to enable an action potential to be recorded from excitable large cells. A third microelectrode may be inserted into the cytoplasm and connected through a further suitable amplifier, etc. to enable the p.d. of the plasmalemma and tonoplast to be studied separately.

To examine the excitatory process in more detail, the voltage clamping technique was adopted from animal cell biophysics by G. P. Findlay and U. Kishimoto, independently, at about the same time (1961). In voltage clamping, the membrane p.d. is controlled by an amplifier. The amplifier senses the difference between the actual membrane potential and the desired value, and the amplified output is used to pass a correcting current through the membrane.

Figure 1.4 shows such a circuit. The amplifier 6a is connected so that its input voltage is that of the membrane and the potentiometer 9a in series. The output current from 6a flows through the internal electrode IJ, through the cell and to earth via the external current electrode EJ. The current flowing across the membrane is recorded at 7' as before while another amplifier 6b and recorder 7 may be used to monitor the actual membrane p.d. The potentiometer 9a must be capable of producing a precise, steady voltage output, together with accurate rectangular pulses and triangular ramps as required.

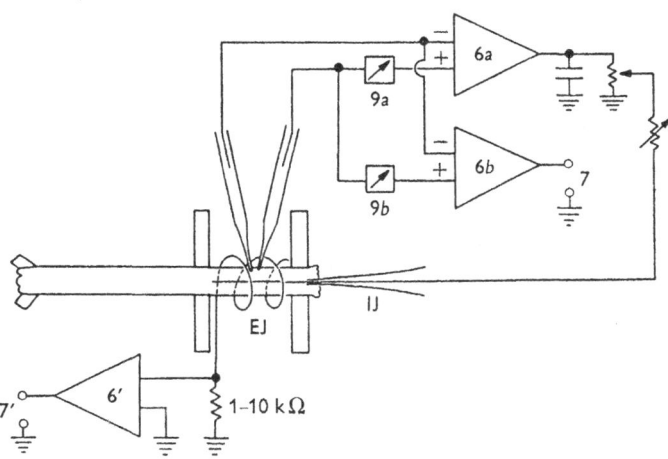

Fig. 1.4 Arrangement for voltage clamping part of the length of a charophyte cell. EJ, IJ are the external and internal current electrodes respectively.

The long wire electrode IJ is the secret of space clamping, i.e. of controlling the membrane p.d. over an appreciable length of the cell. It is possible to make a suitable electrode from platinum/iridium wire, etched with NaCN/NaOH and sealed into tapered glass tubing (see Findlay & Hope, 1964). If the clamp current is introduced into the cell via a micro-electrode, the current is liable to be limited in magnitude and, more important, only 'point clamping' is possible, since the current through the membrane at some distance from the clamp electrode is reduced (in cylindrical cells) by the longitudinal resistance of the vacuolar sap. This effect is much more important during excitation since the membrane conductance then rises sharply: the length constant falls from 15–20 mm to perhaps one-tenth the value, as cable theory shows.

In fig. 1.3, the electrode recording p.d. is shown as 0.42 of the cell half-length from the centrally placed current electrode. Hogg, Williams & Johnston (1968a) showed that with this arrangement, notwithstanding cable effects, the p.d. recorded during a subthreshold pulse of current would be the average for the whole membrane area. No such complication exists in the clamping of the membrane p.d. of a spherical cell such as *Valonia*.

The clamping method was extended when Findlay (1964) controlled the p.d. of the plasmalemma of *C. corallina*, rather than that of the plasmalemma and tonoplast in series, which had been difficult to interpret. This control of the plasmalemma was achieved by inserting the p.d.-recording electrode into the cytoplasm. Then, to obviate many short, disconnected observations obtained by separate re-insertions following 'sealing', the clamped p.d. was swept quickly over a range during the peak of excitability, so that the current:voltage characteristics were obtained for the plasmalemma in the excited state (see, for example, fig. 8.3). Since both plasmalemma and tonoplast exhibit excited states, it is obviously necessary to measure their properties separately.

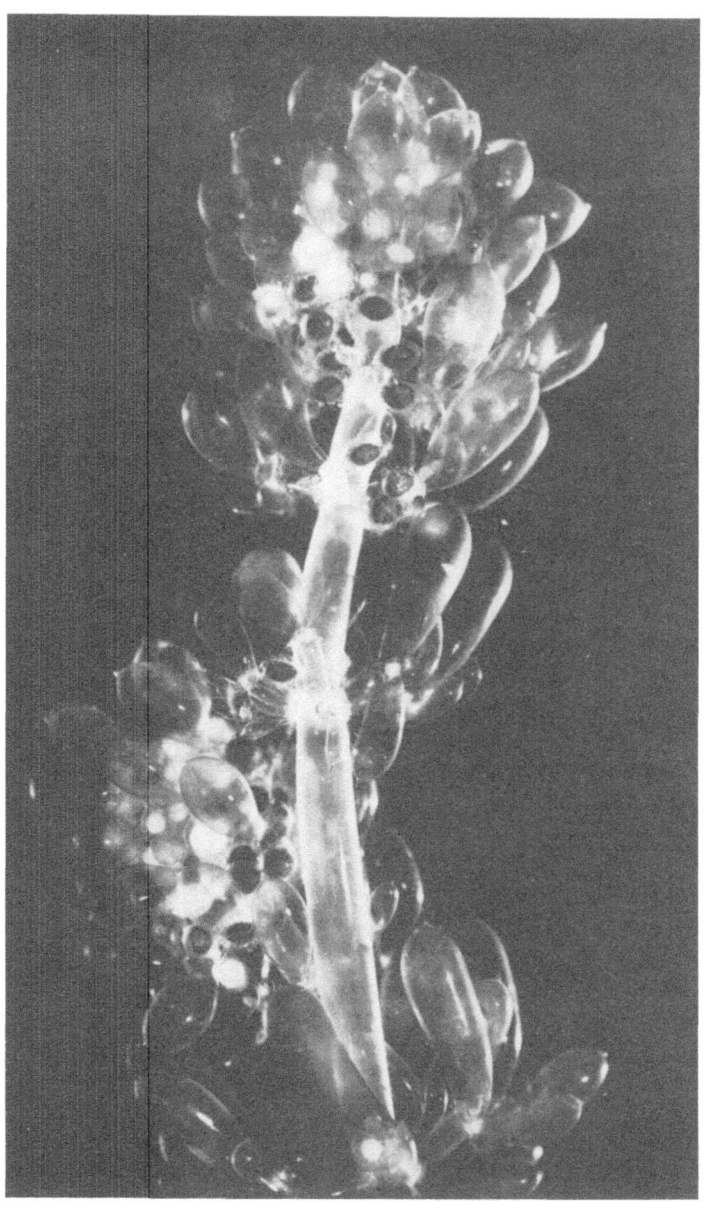

Plate 1 Apical shoot of *Nitella clavata*, showing internodal cells, leaf or whorl cells, and oogonia: cultured material. (Photo: B. T. Lester.)

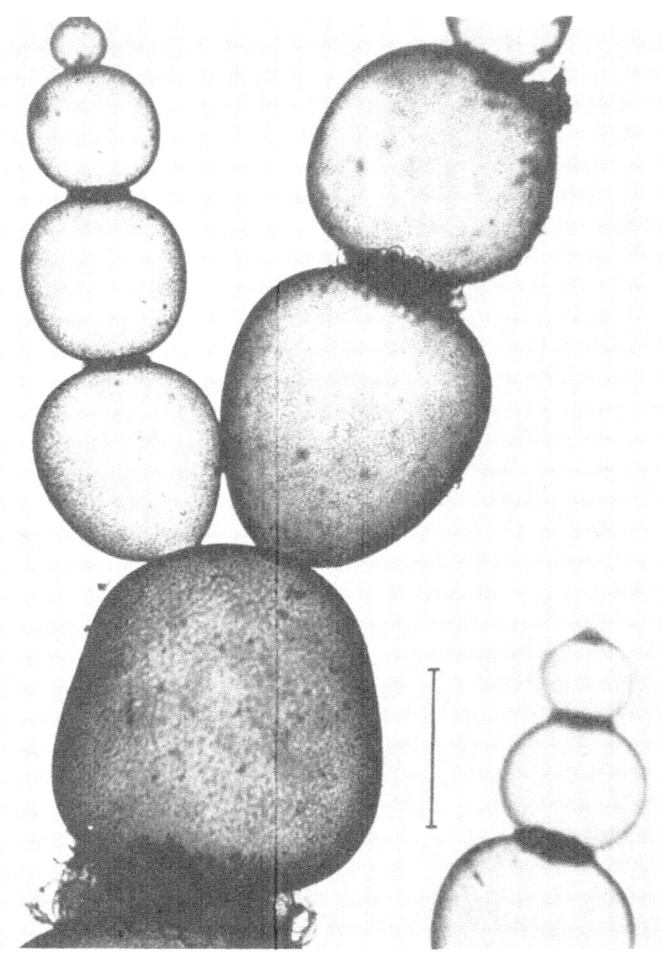

Plate 2   Cells of the red alga *Griffithsia* sp. Collected material.
The scale mark represents 1 mm. (Photo: B. T. Lester.)

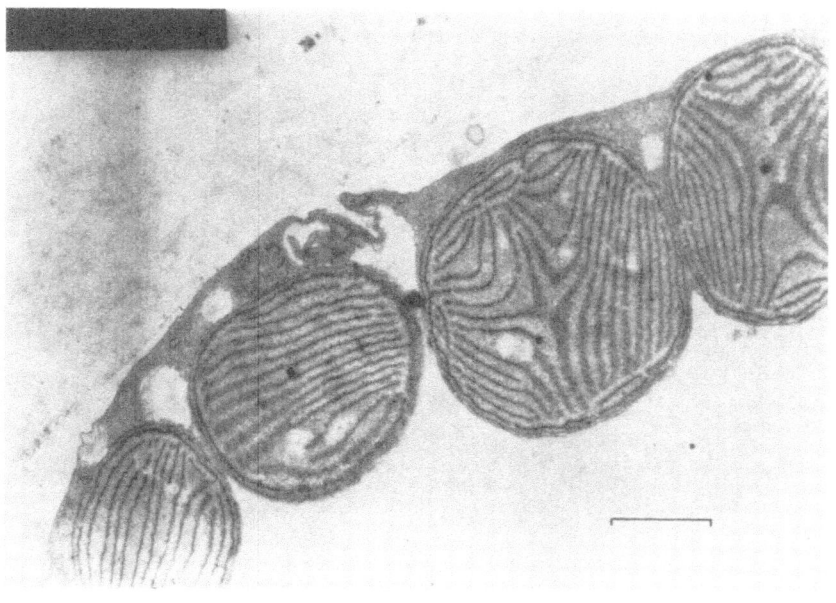

Plate 3 Electron micrograph of a section through a cell of *Griffithsia* showing chloroplasts, a mitochondrion, cell wall and thin cytoplasm. The scale mark represents 1 μm. (Electron micrograph from A. W. D. Larkum and Sydney University E.M. unit.)

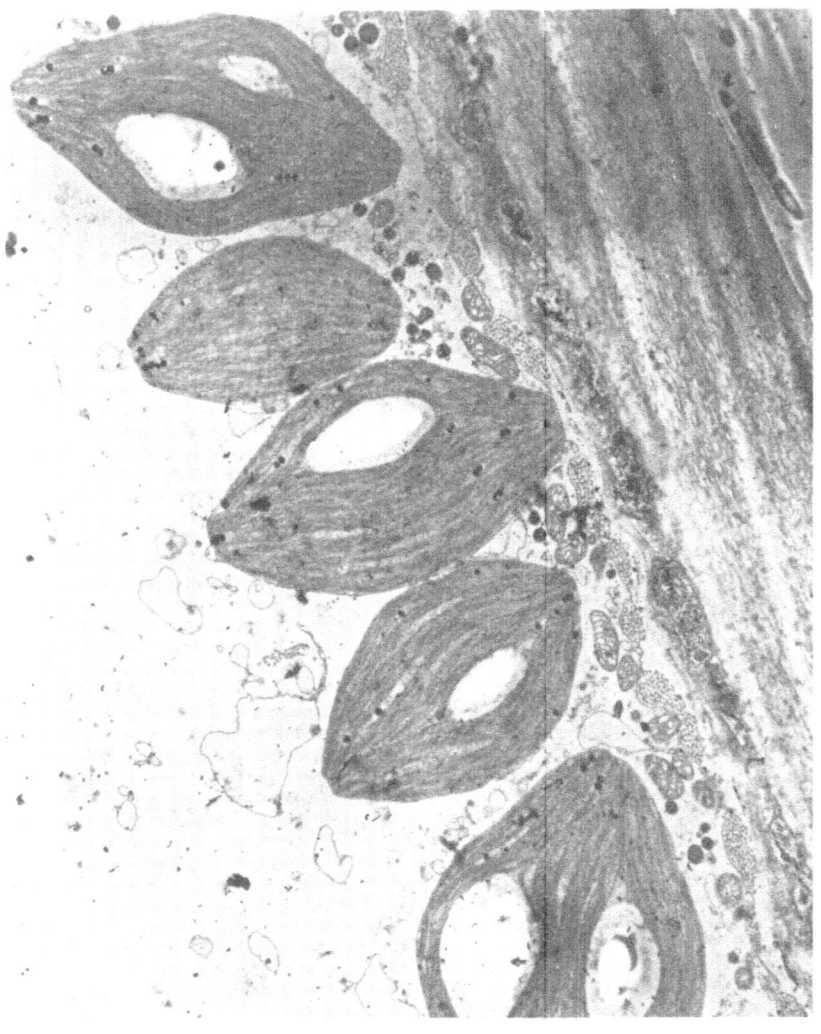

Plate 4   Electron micrograph of a section of an internodal cell of *Chara corallina*. Fixed in 3% glutaraldehyde, stained in 2% osmium tetroxide. Note the row of membrane organelles at the plasmalemma. (Electron micrograph from M. Vesk and Sydney University E.M. unit.)

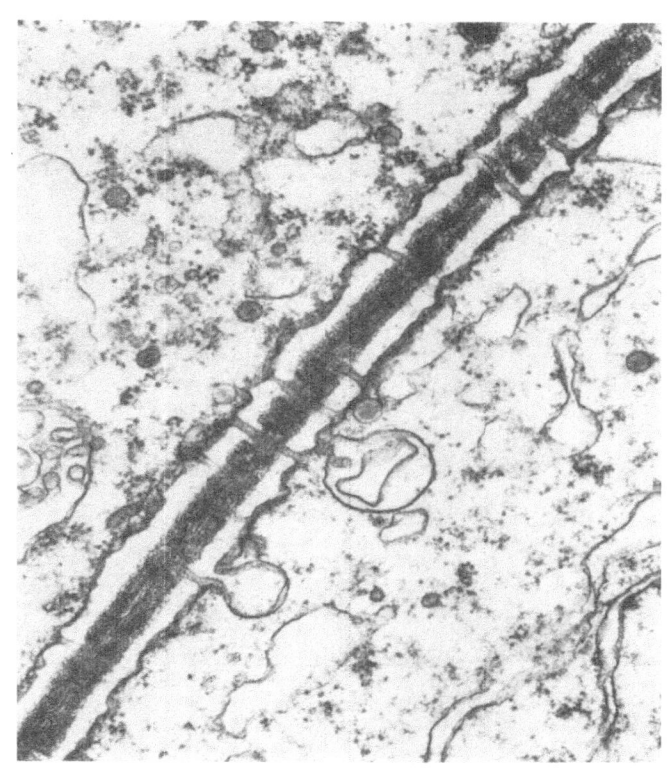

(a)

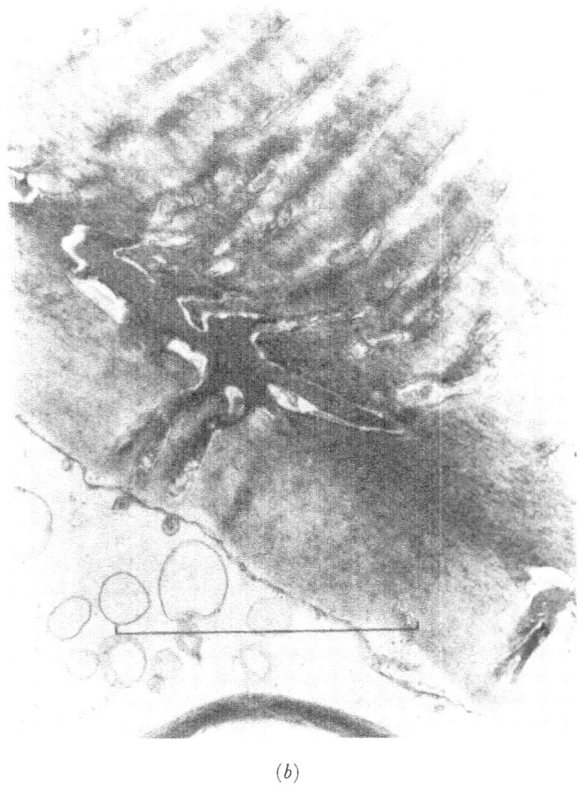

(b)

Plate 5 (a) Plasmodesmata and cell wall between two nodal cells of *Chara corallina*. (Electron micrograph from J. D. Pickett-Heaps.) (b) The cell wall and plasmodesmata between a nodal and an internodal cell in *C. corallina*. The scale mark represents 5 μm. (Electron micrograph from T. E. Bostrom and Sydney University E.M. unit.)

CHAPTER 2

Two hundred years: early physiology

Giant algal cells appear in science with Corti (1774, 1776) who observed and reported protoplasmic streaming in charophyte cells long before protoplasm existed as a concept. By the early years of the next century these observations were being repeated, and the small stream of publications that arose has continued until the present day. For most of this period the motion of charophyte protoplasm was the only feature of giant algal cells which attracted the physically-minded physiologist.

Many in this period studied the effects of physical variables upon the rate of movement: for example the measurements of the effect of temperature make a considerable list (Corti, 1774, 1776; Dutrochet, 1837; Nägeli, 1860; Cohn, 1871; Velten, 1876; Ewart, 1903; Lambers, 1925; Romijn, 1931; Umrath, 1934; Hayashi, 1960). It is tempting to dismiss all these works except the last, for only Hayashi made progress with what seems now to be the first question to be asked: how does temperature affect the two quantities that determine the speed, i.e. the viscosity of the protoplasm and the motive force? The earlier workers however did not all ignore the need to provide explanations for the phenomena they observed. In particular Ewart (1903) did not feel that it was enough to observe physical effects on a biological phenomenon. His book is an elegant physical analysis of charophyte streaming; of its division into chapters headed 'Physics and Chemistry' and 'Physiology', he says: 'Under the first heading those phenomena of protoplasmic movement are discussed which can be directly referred to physical and chemical causes, whereas under "Physiology" we deal with those "vital" phenomena which cannot as yet be thus resolved.' He published the first velocity distribution of charophyte streaming, and used it to deduce the site of the motive force; and calculated the expenditure of energy by a charophyte cell in streaming, using the observed speed and the viscosity of albumen solution, which he measured

19

B

himself. The answer was higher than that provided by modern work, by a factor of 1.5.

His work illustrates well the enduring utility in physiology of concepts with a physical basis; it shows up, by contrast, the evanescence of experiments based on little more than the desire to observe, still more of those based on merely 'physiological' concepts.

By Ewart's time, the major genera and species of giant algal cells were known, although the charophytes were the commonest of this experimental material, both for their interest and because of their distribution throughout Europe. *Valonia*, *Halicystis* and *Acetabularia*, available to Europeans only at Naples, were comparatively neglected. The problems that were to make these marine genera (together with the charophytes) into popular experimental material were, however, currently under discussion – they were membrane permeability problems.

At that time Pfeffer (1877) had already published his classical work on osmosis, Nägeli (1855) had long since observed plasmolysis, and de Vries (1884) had studied the turgor of strips of plant tissue in solutions of different concentration. Nägeli had also observed that extruded protoplasmic drops were surrounded by membrane. The view that has since remained the orthodox dogma, that the living cell has its watery protoplasm separated from its watery environments by membranes, was by then well established. Both plasmalemma and tonoplast were recognised as diffusion barriers (Pfeffer, 1877), and the permeability properties of membranes were attributed by the same author to both the affinity of the membrane material for solute and the properties of their aqueous pores, if any. This grand generalisation became more specific and more testable, with the work of Overton, and became general again in a different sense. Overton, beginning with an interest in the penetration of ethanol into cells, took up the problem of selective permeability. After studies of penetration of 500 different compounds, he was able to connect the results with a physically quantifiable concept, the oil solubility of the solute (Overton, 1899) and with the (in principle) testable hypothesis that membranes contained lipids. He was able to show that these ideas could be applied to a wide variety of cells, including algae, root hairs, erythrocytes and then muscle cells (Overton, 1900, 1902). Overton was strongly opposed by

Traube, who was, however, not able to find a physical basis for his own concept of retention pressure of membrane for solute. Membrane permeability was thus, at the turn of the century, a topic of debate and of enthusiastic work; and the electrical aspects of the question were being tackled too. Stewart (1899), Bugarsky & Tangl (1898), and Fraenckel (1904) were studying the electric conductivity of red blood cell suspensions, showing in effect that the cell membrane had a high electrical resistance. Electrical resistance measurements were to be for many years separate things from electrical potential difference measurements – Blinks perhaps was to be the first to relate them. The turn of the century did see some potential measurements – after the pioneering work of Haake (1892). Hörmann (1898) studied the effect of electric shocks on streaming and found the charophyte action potential. The finding that chloroform would kill sections of cell membrane was a small technical point of some value to later workers.

The large volumes of sap contained by marine coenocytes made them useful for chemical analysis in times when techniques were not especially accurate in small-scale work, but their use developed slowly. Perhaps concepts were rate-limiting rather than techniques: at all events Meyer (1891) and Hansen (1893) each worked on *Valonia* for a summer at Naples, and both showed differences in ionic composition between sap and sea water. Nathansohn (1903) soaked *Codium* cells for 2–10 days in sodium nitrate solutions, and found that the nitrate ion never came to a concentration in the sap equalling that outside:

t/day	$[NaNO_3]_o$/mM	$[NaNO_3]_i$/mM	$[NaNO_3]_i$/$[NaNO_3]_o$	ψ_{NO_3}/mV
2	59	34	0.58	−14
4	59	33	0.56	−14.5
10	59	40	0.68	−9.7
5	118	52	0.44	−20.6
5	118	58	0.49	−18
2	447	192	0.43	−21
2	565 (recalculated)	232	0.41	−22.4

This went no further, Nathansohn turning to flowering plants for future studies; the next worker to use the attractively large coenocytes of *Valonia* was Wodehouse (1917) from Osterhout's laboratory. He was conscious of the use that might be made of *Valonia* in ion permeability studies, but did not himself tread the path very far.

A few years later Gardner suggested using *Nitella clavata*, for experiments on salt uptake, to Hoagland & Davis (1923) at Berkeley. Osterhout (1922) had already started work on *Nitella flexilis* from a local pond in Cambridge, Mass., and he and his students were studying exosmosis, permeability (Irwin, 1923) and electrical conductivity.

His early electrical work on *Nitella* did not go far – nor indeed did the first microelectrode work or most of the early radioactive tracer work. Clearly techniques are not to be judged on their first appearance.

Electrical measurements at this time often centred around the idea that the electrical conductivity of cells is related to the integrity of the cell membrane. This idea underlies work already quoted on red blood cell suspensions, and the work of Osterhout (1914, 1920) on the conductivity of *Laminaria* fronds and of S. C. Brooks (1917) on the same material. Since in multicellular tissues and suspensions it is not always obvious how to distinguish the parallel resistances of cell interior and outside medium, the idea suggested itself that the cell be bathed in a medium of conductivity similar to that of its interior. If the electric resistance were observed to be higher than that of the medium, this could then be attributed to the impermeability of the cell membrane. This idea was so persuasive that we find Remington (1928) still trying to use the same method with beet slices. Osterhout's first electrical measurement on giant cells (1922) used this method; fig. 2.1 reproduces his diagram of the set-up. A bundle of *Nitella* cells lay in the chamber between two electrodes, in seawater diluted with three parts of distilled water to match the conductivity of the sap. Nothing about the apparatus suggests that we are dealing with a new and exciting material; the *Nitella* cells could have been replaced by pieces of multicellular tissue with no change in the set-up or results. Several years elapse before there appear any electrical measurements that exploit the characteristics of the charophyte cell. This point is perhaps near a watershed in Osterhout's career. Previously he had been much occupied with the concepts of injury and death, which refused to be usefully quantified: in fact they were a dead end. But soon new ideas were to catch his attention, and he and his collaborators were to keep algal cells in the forefront of physiology for some years.

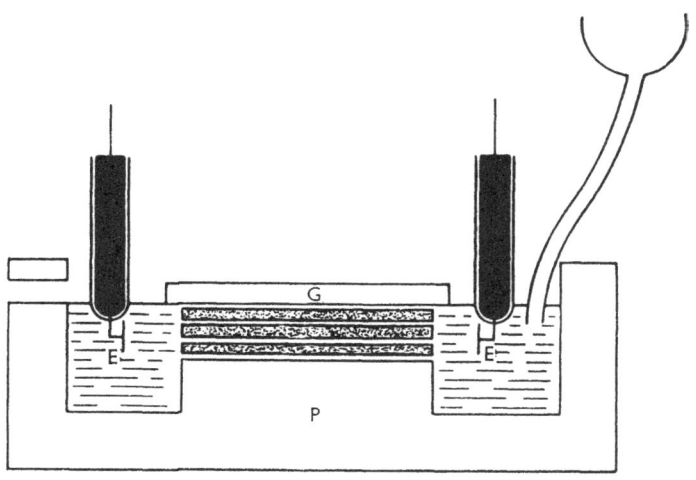

Fig. 2.1 Apparatus for measuring the conductivity of *Nitella*. The cells are placed in a trough in a block of paraffin wax P and covered with a glass plate G. E = platinum electrode (from Osterhout, 1922).

The early twenties was in fact an exciting time for new developments. In Europe there had been a lull in publications from 1914–21, but in America the new *Journal of general Physiology* had been started and was full of good American work. Often the editors seemed intent on filling each number themselves. Loeb, incredibly prolific, with a team of several research assistants who made the measurements, published two or three papers of some intellectual quality in each number; and Osterhout, prolific with short papers, apparently made his own measurements, since his students, once paternally introduced (Osterhout, 1918), published under their own names (Marian Irwin, Matilda Brooks, Helen S. Thomas and F. G. Gustafson).

The period from 1917 to 1940 saw many measurements made for the first time on giant algal cells; a simple list makes the point.

1917 Wodehouse made the first study explicitly related to permeability, of ions in sap of *Valonia* cells. Not quantitative.
1918 The first volume of the *Journal of general Physiology* appeared.
1919 Crozier published the first measurements of pH of *Valonia* sap. The value found was 5.9, while the external seawater was at 8.2.

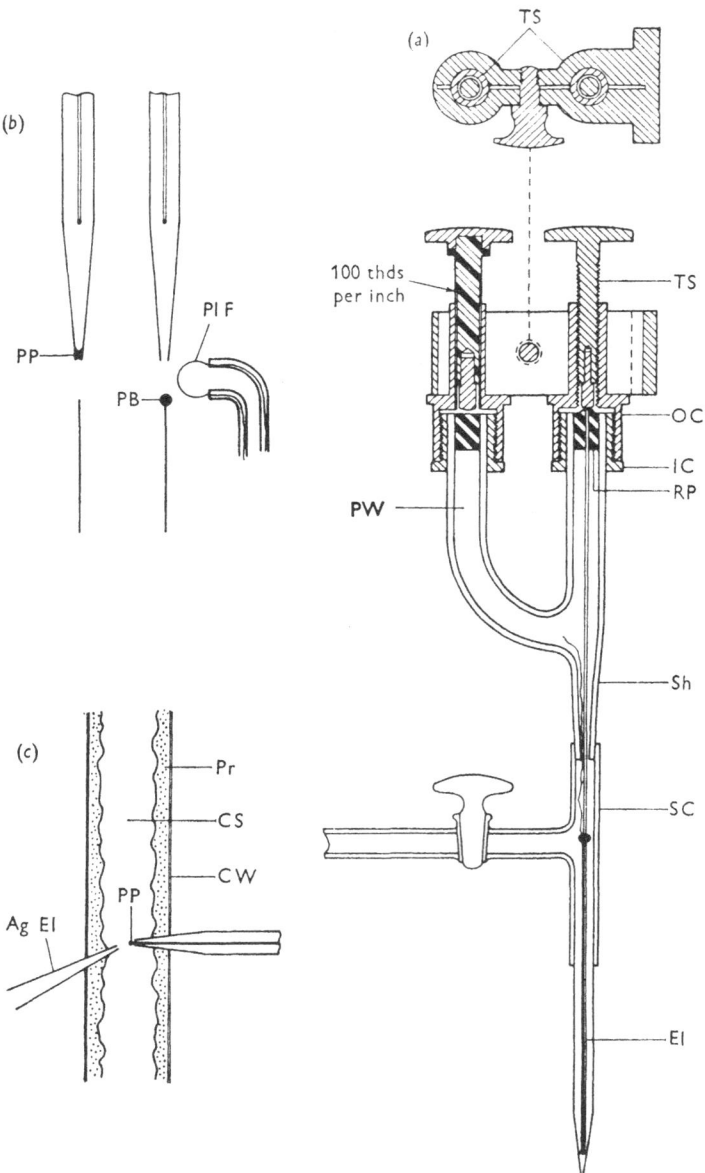

Fig. 2.2 The microelectrode used by Taylor & Whitaker (1928) to measure the pH of the vacuolar sap of *Nitella*. (*a*) Microelectrode enclosed within its quartz sheath which is terminal part of the micropipette apparatus. El, microelectrode; IC, inner cap; OC, outer cap; PW, platinum wire; RP, rubber plug enclosing

24

1922 Osterhout published the results of qualitative tests for the penetration of nitrate into *Nitella* cells, and the first conductivity measurements on *Nitella*.

1922 M. M. Brooks reported qualitative measurements of cation penetration into *Nitella*.

1923 Irwin made quantitative studies on dye penetration into *Nitella* as outside pH was varied; these were the first quantitative measurements of the kind.

1923 Hoagland & Davis described the first measurements of ion concentrations and pH in *Nitella* sap – they found the values listed in table 5.1, and an internal pH of 5.2 after several days immersion of the cells in solutions ranging in pH from 8 to 5.

1926 Taylor & Whitaker reported the first insertion of microelectrodes into *Valonia*. They used a sensitive galvanometer and calculated the membrane potential difference from the current and the measured resistance of the microelectrodes. The result, $\psi_{vo} = +2$ mV, was of the right sign, but probably too small in magnitude.

1926 Irwin found the influx of dye to be proportional to the concentration of neutral dye molecules in external solution.

1927–8 Osterhout, Damon, Jacques and Harris reported the 'asymmetry of the inner and outer protoplasmic surfaces' in *Valonia* and in *Nitella*. This observation, of an electric p.d. across the protoplasm when its faces were bathed in similar solutions, moulded American thought on p.d.s. in giant algal cells for two and a half decades. It is an interesting comparison with Ussing's shortcircuit technique.

1927–8 Gelfan inserted two microelectrodes into *Nitella* cells and made conductivity measurements in sap and cytoplasm.

1927 Jost made the first measurements of the p.d. between inside and outside of a cell of *Nitella* using an inserted electrode; he found ψ_{vo} to be -46 mV.

---

small brass tube which supports quartz capillary on which is cemented electrode; SC, stop-cock; Sh, shank of micropipette which is filled with mercury; TS, thumb-screw. (*b*) Two views of micropipette tip, showing how it is sealed with paraffin plug. PB, paraffin ball on tip of quartz needle; Pl F, platinum filament heated electrically for melting paraffin ball whereupon tip of pipette is brought in contact with the melting paraffin; PP, paraffin plug, formerly paraffin ball, in tip of pipette. (*c*) Diagram of portion of cell of *Nitella* showing hydrogen electrode and agar-KCl electrode in cell-sap. Ag El, agar-KCl electrode; CS, cell-sap; CW, cell wall; PP, paraffin plug having been pushed out by tip of hydrogen electrode; Pr, streaming protoplasm, granular in contrast to the clear cell-sap.

1928   Taylor & Whitaker reported the first *in situ* measurement of the pH of the sap of a cell, using *Nitella* (fig. 2.2). Their value, 5.1, agreed with that of Hoagland & Davis (1923 – see Table 5.2) and those of later workers.

1928   Gicklhorn & Umrath published the results of insertions of microelectrodes, of 8 μm tip diameter, into a number of plant cells. Three p.d. measurements were made on *Nitella* cells. The measured potentials were negative inside but of small magnitudes (-3 to -19 mV) which suggests injured cells. A thermionic valve amplifier typical of the period was used (fig. 2.3).

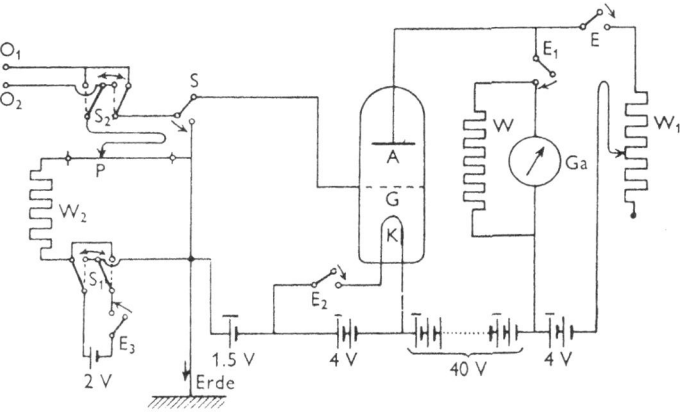

Fig. 2.3   The circuit and connections of the electrometer millivoltmeter used by Gicklhorn & Umrath (1928) for the measurement of intracellular potentials in *Nitella*. P, potentiometer; S, switch; O_1 and O_2, inputs; A, anode; G, valve grid; K, cathode; E, current keys; W, resistance; Erde, earth.

1928   Brooks & Gelfan inserted microelectrodes, of 5 μm tip diameter, into *Nitella* leaf cells. A low potential was found, which showed a positive correlation with cell size, perhaps because of the low impedance of the measuring circuit. The polarity was stated to be *positive* inside.

1929   Lark-Horovitz made the first radioactive tracer measurements on *Valonia* in Osterhout's laboratory. The living cell appeared permeable to radon and impermeable to lead ions, to which the dead cell was permeable. The original article appears as fig. 2.4. These results tended to confirm the then current picture of the permeability of membranes to neutral molecules but not to ions.

aero-engine to below 7 : 1 ; hence there is a distinct loss of possible efficiency. Diisobutylene or diamylene, having better anti-knock properties than benzol, could permit of a higher H.U.C.R., and, moreover, blends of these hydrocarbons would not be liable to freeze at high altitudes, both substances being liquid at - 45° C. in an undiluted state. Diisobutylene may be conveniently prepared by the polymerisation of the isobutylene content of 'cracked' gases by means of sulphuric acid, while diamylene may be obtained by the similar treatment of either trimethyl ethylene or tertiary amyl alcohol.

Since this work was completed it has been found that E.P. 253,131 covers the use of these two olefines, among others, as anti-knock dopes, and describes them as being better than benzol for this purpose, but no comparative figures are quoted.

It is interesting to note that of the olefines we have tested, those which are the more stable towards bromine, sulphuric acid, potassium permanganate, and potassium bichromate, are the more effective in suppressing knocking.
A. W. NASH.
DONALD A. HOWES.
Department of Oil Engineering and Refining,
University of Birmingham.

A Permeability Test with Radioactive Indicators.

CERTAIN investigators (see, for example, W. J. V. Osterhout, "Some Fundamental Problems in Cellular Physiology," 1927 ; especially pages 36-48) believe that the protoplasm of the living cell is permeable only to undissociated molecules but impermeable to ions.

It seemed possible to me to test this theory with the method of radioactive indicators [1] (Hevesy-Paneth). The advantage of this method is that only very small amounts of the ions which enter the cell are necessary and that a very small concentration can be detected. Radioactive lead (thorium-B) was used as an indicator for lead ions, and therefore lead nitrate was dissolved in sea water so as to make it 10^{-8}–10^{-9} M in respect to lead ions. Cells of *Valonia macrophysa* were used since the large volume and the amount of sap available make the investigation easier, and since investigations of the permeability of this cell were carried out by Osterhout and his collaborators.

To test whether or not the presence of lead causes any injury to the cell, the cells were placed in sea water with different amounts of lead nitrate added, and for several months the behaviour of the cells observed. The cells did not change in colour or rigidity, and were, according to Dr. L. R. Blinks, who kept them in the same laboratory with other cells, in a normal state, judged from macroscopic appearance.

For the permeability experiments, the cells were placed *in sea water containing a known amount of lead nitrate and thorium-B*. After 20 or 30 hours the cells were taken out, washed off with inactive sea water, and dried on blotting paper. The sap was removed, a certain amount (0.2–0.3 c.c.) evaporated in a watch glass, and the radioactivity measured in an α-ray electroscope. The activity of the same amount of the original solution and of the sea water in which the cells were kept was measured. In this way we ascertained how much lead is absorbed by the cell wall and how much enters the vacuole. In all experiments (14 cells) it was found that about 50 per cent of the lead ions present in the original solutions are absorbed by the cell wall, but that practically no lead

enters the vacuole.[2] The same experiments were carried out with cells which had been kept in sea water plus lead nitrate for four months. Also in this case no lead could be found in the vacuole.

One may conclude that all the lead which disappears from the sea water is adsorbed by the cell wall or the protoplasm forming an insoluble compound which cannot enter the vacuole. In this case one would expect that in dead cells also the lead would be fixed at the cell walls and therefore cannot be found in the sap. Experiments with three dead cells have shown that lead does enter a dead cell. It is apparently fixed there to small particles of organic matter which are to be found always in dead cells. Therefore it cannot diffuse back into the surrounding sea water and *an apparent concentration* of lead in the dead cells takes place.

It was interesting to see whether radium emanation, being a rare gas, would enter the cells, as one would expect from the theory. Small capillaries (16 mm. long), filled with radium emanation (about 0.01 m.c.), were broken under the sea water containing the cells to be tested. It was found that already after one hour the sap is approximately as active as the surrounding sea water (15 cells were investigated).

After every experiment, Dr. L. R. Blinks examined the macroscopic appearance of the cells and tested the sap for sulphate ions. (The presence of sulphate ions in our lead experiments and the sap of every single cell in the experiments with radium emanation was tested in this way. Injury was found in one cell out of a total of three, exposed for 20 hours in radium emanation, and traces of sulphate ions in two cases out of twelve, after 1 to 2 hours exposure in radium emanation. One cell that had been in lead nitrate for four months was soft, but did not give any sulphate reaction and did not show any sign of injury in our test.

Summary.—Using radioactive indicators for testing the permeability of single cells of *Valonia macrophysa*, it was found that lead ions do not enter the sap of the living cell even if the cells are kept for several months in lead nitrate solution. Lead ions enter readily the sap of dead cells. Radium emanation, being a rare gas, is already after one hour distributed evenly between the cell sap of living cells and the surrounding sea water.

This investigation was carried out in the spring of 1927 during our stay at the Rockefeller Institute for Medical Research, New York City, and we are indebted to the International Education Board who made our stay at the Rockefeller Institute possible.
KARL LARK-HOROVITZ.
Physics Department,
Purdue University,
Indiana.

Molecular Constants of Hydrogen.

ONE of us recently published a table of constants for the neutral hydrogen molecule (*Proc. Nat. Acad. Sci.*, 14. 12 ; 1928). The most uncertain quantity in that table was the value of the moment of inertia for the '*B*' level. The value given (1.99 × 10⁻⁴⁰) is based on Hori's very doubtful interpretation of Witmer's band progression B_3–A_n. We have now photographed the entire B–A system in the second order of a ten-foot vacuum spectrograph, designed by Prof. J. J. Hopfield

---

[1] That is, to determine the amount of ions present, of a certain kind, by the determination of the radioactive isotope mixed with them. Since a chemical separation of isotopes is impossible, the change in activity of the radioactive isotopes is the indicator for changes in the concentration of the inactive ion.

[2] A trace of activity which was found twice immediately after drying is due to traces of thorium-C. This may have entered the cell in ionic form, but since thorium-C is present only in an extremely small concentration, this is not contradictory to any other experiment on permeability. Such a small amount may possibly also enter in other cases, but could not be detected. On the other hand, thorium-C shows in neutral solutions a quasi-colloidal behaviour and may have entered the cell in form of an undissociated complex.

No. 3095, Vol. 123]

Fig. 2.4   Reproduction of the article by Lark-Horovitz (1929) reporting the use of the radioactive tracers Thorium-B and Radon to test for permeation into cells of *Valonia*.

1929 Blinks, Harris & Osterhout reported *Nitella* action potentials and interpreted them in terms of the local circuit theory of propagation. A thermionic valve amplifier was used, with a string galvanometer.

1929 Osterhout & Harris made measurements of changes of membrane p.d. with salt concentration. They used *Nitella* cells with external wick contacts (fig. 2.5), and interpreted the results in terms of diffusion potentials. From here on Osterhout used much more quantifiable concepts to discuss his results.

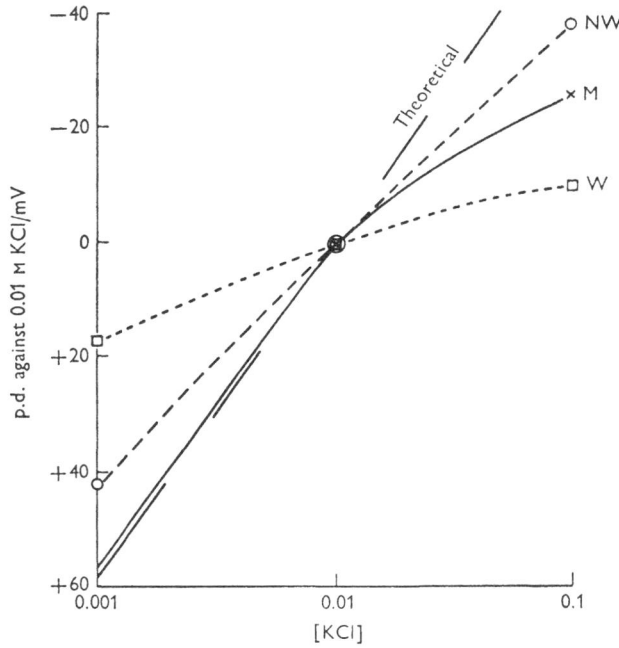

Fig. 2.5   Results of Osterhout & Harris (1929) showing the effect of variation in the concentration of KCl in the medium on the p.d. between two ends of a *Nitella* cell, one end kept in 10 mM KCl. 'Theoretical' referred to the expectation if the p.d. were a diffusion potential with the mobility of the chloride ion negligibly small; W to experiments with cell walls; NW to intact cells, and M to the inferred concentration effect on the protoplasm alone.

1930 Blinks made the first measurements, in *Nitella* and *Valonia*, of the d.c. resistance of plant cells, and of the variation of resistance with current (1930*a*, 1930*b*).

1930   Umrath reported the first studies with inserted electrodes of the action potential in *Nitella*, using a Lindemann electrometer, which served him for measuring and recording for many years.

1930   Collander's first 'Permeabilitätsstudien an *Chara ceratophylla*' appeared. It was first of three, and it dealt with the composition of the cell sap.

1933   Collander & Bärlund published the most complete study on nonelectrolyte permeation in a single cell. Overton's ideas about membrane permeability were put on a quantitative basis, using the idea of correlating permeability with oil:water partition coefficients.

1937–8   Cole & Curtis measured transverse a.c. impedance of *Nitella* at rest and during the action potential.

From this point on, innovations were few, and the protagonists already on the scene continued to speak their lines more or less unaffected by the presence of the others. This is at any rate true until the renaissance of interest in the membrane properties of giant algal cells that occurred during the 1950s and 1960s. The revival owed much to the new availability of radioactive isotopes of many elements.

During the two decades 1930–50 centres of giant cell work were at Collander's laboratory in Helsinki, Umrath's at Graz, Osterhout's in New York, Brooks' at Berkeley, Hoagland's at Davis. Their isolation is notable. Obviously the Americans knew each other, but their lists of references generally stop at the Atlantic coast. At first, in the early thirties, Umrath was reading Osterhout's work, abstracting it for *Protoplasma* and quoting it. Osterhout seems not to have taken any notice of Umrath's work after 1931.

The irony of this requires, for its appreciation, some examination of their work. Osterhout, and his collaborators Hill, Harris and Blinks, observed that the potential difference across the protoplasmic layer could be regarded as the sum of two membrane p.d.s, one at each protoplasmic surface. Characteristically, Osterhout named these surfaces 'X' and 'Y', instead of 'plasmalemma' and 'tonoplast'. They showed that the resting potential of the *Nitella* internode (ψ_{vo}) was essentially a potassium diffusion potential, and then found that it was possible to obtain cells with high (normal) values of ψ_{vo} which were, however, not depolarised by increasing the outside potassium concentration. (This is possible still.) The observation was explained by the hypothesis that X (the outer layer) had become insensitive to potassium (P_K was lower) while Y had the normal sensitivity to potassium concentration – hence ψ_{vo} was developed across Y, and hence the potassium concentration in the cytoplasm was low. In experiments on effects of toxic agents on the value of ψ_{vo} it was often hypothesised that X or Y was the first to be affected, according to the time course of the potential changes.

The technique needed to test Osterhout's hypotheses about X and Y was the insertion of a microelectrode into the cytoplasm, Osterhout's 'W'. This was just what Umrath had been doing since 1930. His first paper on microelectrode techniques (1930) is quoted by Osterhout (1931), but a footnote in the same review offers Osterhout's opinion that 'a capillary pushed into protoplasm may not break the surface film but may form a deep pocket (Chambers (1922)). Even if the film be broken a new one will form so that it seems impossible to get inside the protoplasmic surface unless possibly at the instant when liquid is actually flowing out of the capillary.' This quite rational belief may have been all that stood in the way of rapid progress with the microelectrode technique.

It is clear that Osterhout and his collaborators made a major contribution to the electrophysiology of giant algal cells, and indeed to physiology. The paper in 1928 by Osterhout & Harris pioneered the quantitative study of membrane potential differences in cells. Osterhout was always full of ideas: he for example first suggested carrier molecules, and his proliferating papers show still an urge to think and to communicate the ideas that flowed.

His colleague Blinks can be said, with Osterhout, to have laid the foundation for present-day studies of giant algal cells. Blinks' review (1949) summed up the work of three decades, and within the limits of the techniques adopted, took knowledge of giant algal cell membranes as far as it could then go. If Blinks too made much of the separate properties of X and Y, it must be noted that even now, most marine genera cannot readily be stuck with a microelectrode in such a way as to measure ψ_{co} and ψ_{vc} separately. Within this limitation he investigated membrane potentials, using the Nernst and Henderson equations as a basis for quantifying them; he studied effects of light, ammonia, anoxia and phenolic substances. He related membrane conductances to ionic mobility (especially K^+ mobility). His investigation with Pickett stood for many years as the only evidence for rejecting the idea of redox potential differences producing (or affecting) ψ_{vo}. At about the time of the 1949 review, Blinks turned to photosynthesis, with equally fortunate results for plant physiology.

The work of Osterhout's contemporary, Collander, also causes one to ponder on the personal factors which determine the quality of an author's work. Like Loeb, who transmitted his ideas of 'antagonism' to Osterhout, Collander seems early in his career to have become interested in applying physical chemistry in biology. His early papers were concerned with the permeability of both artificial membranes – copper ferricyanide, gelatin or collodion – and biological membranes. By 1930 he was publishing analyses of the sap of *Chara ceratophylla*. At about this time he must have made the important decision to leave the question of ionic permeability and to work instead with nonelectrolytes, still using *C. ceratophylla*. This was the turning point, for it provided him with data which, if they could not yet be quantitatively explained from molecular mechanisms, could at least be meaningfully compared with chemical data using a membrane model. Just as important as this choice of problem was the fact that Collander proved equal to the task of performing a truly vast number of careful measurements.

The permeabilities of nearly a hundred chemical compounds were determined by efflux measurements, the quantity of each substance being determined by an appropriate chemical method. If solutes crossed the membrane by dissolving in a lipid layer, their permeabilities should be directly related to their oil:water partition coefficients and to their diffusion coefficients in oil. The former is the deciding factor, as we shall see in chapter 4. Collander (1950) measured oil:water partition coefficients for most of the compounds he had used, with olive oil and isobutanol as alternative models of the membrane lipid. The good correlation he found between permeability and partition coefficient put the lipid membrane theory on a quantitative basis. It stands today, enhanced rather than superseded by the work of Diamond & Wright, who have left a brief but enthusiastic appreciation of Collander's work (p. 615 of their review, 1969).

So far, the existence of the surface membrane in the plant cell has been assumed as a central dogma. But this dogma did not survive from the nineteenth century without challenge. In 1915 we find Fischer suggesting that the phenomena observed to that date did not require the existence of a surface membrane to the cell. Though this was perhaps not a serious challenge to the dogma, it was sufficiently strong for Remington (1928) to point out in a careful review that the body of the evidence was against Fischer.

Years later the question arose again, and a number of different workers brought independent evidence to show that the outer surface of the plant cell appeared to be quite permeable to small solutes. Always in the background, the principle attributed to William of Ockham (1487) required that there be postulated only the minimum number of bounding membranes. The plant cell already had a tonoplast, there for all to see (plate 6). Need another be postulated at the boundary of the protoplast? The osmotic behaviour of vacuolated plant cells could readily be explained by the semi-permeability of the tonoplast. The unexpectedly small fraction of mature plant tissue occupied by the apparent osmotic volume led Briggs and his students at Cambridge quietly to abandon the concept of a plasmalemma in plant cells. Robertson & Turner (1945), studying salt respiration in storage tissue, found no effect on respiration of salt in the vacuole, but a prompt increase in respiration occurred if the same salt were added to the outside solution. It was natural to interpret this in the terms being propounded by Briggs. Hope (1951), working with bean roots, found long drifts of potential when the outside solution was changed, and then large movements of salt into and out of the tissue when the external concentration was raised or lowered (Hope, 1953). This seemed evidence for the ready movement of salt into and out of the protoplasm, which was thought to be the only structure which could exhibit so much ion exchange. Hope & Robertson (1953) then proceeded to examine the literature on giant algal cells and found, it need not surprise the reader of this chapter, nothing in the work of Osterhout, Blinks, Cole & Curtis, or Collander that necessitated any definite properties being assigned to the plasmalemma. In abolishing the concept Hope & Robertson did perhaps less than justice to the microdissection work of Plowe (1931): but

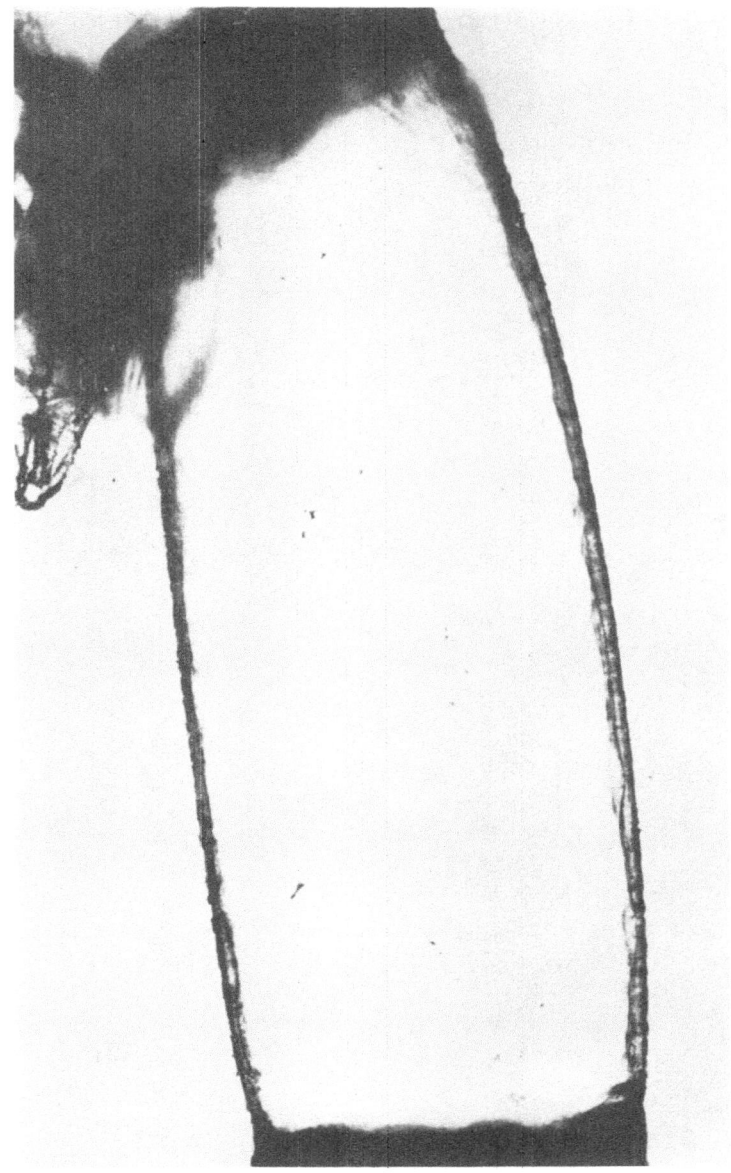

Plate 6 Young internodal cell of *Chara corallina*. Note expecially the clear phase boundary between the cytoplasm and the vacuole. (Photomicrograph: J. P. Fairburn.)

they rightly showed that the biophysical work to that date could be interpreted in terms of Osterhout's Y and a cytoplasm in Donnan equilibrium with the medium.

So for a few years the matter rested. Electron microscopy, then in its infancy, produced pictures of the plant tonoplast, but the osmium tetroxide fixative then in use showed nothing at the cytoplasm–wall boundary. Occam's razor seemed to have fallen on the plasmalemma, at least in the world of plant cells.

Experiments with radioactive tracers, which might have made the matter plain, had not at the time been performed. After Lark-Horovitz's pioneering study in Osterhout's laboratory, nothing more was done for ten years, when Brooks and his student Mullins began such work again (Mullins & Brooks, 1939). This seems to have been the first use of ^{42}K in plant physiology. Soon after, at Collander's instigation, Holm-Jensen, Krogh & Wartiovaara (1944) studied fluxes of potassium in *Nitella*. Neither group seems to have had sufficient resources to do much useful work, though the conclusions of Holm-Jensen *et al.* have been vindicated by later studies.

Microelectrode measurements by Walker (1955) produced two pieces of evidence in favour of the reinstatement of the plasmalemma concept. First, there was found to be a considerable electric resistance between the environment and the cytoplasm of *Nitella*. Secondly, in the presence of calcium in the environment, the potential difference across the cell surface was much too high for calcium to be in electrochemical equilibrium between the two phases, and so a barrier to the diffusion of calcium was indicated.

Several years later these arguments were reinforced by the elegant flux measurements of MacRobbie & Dainty (1958) and then of Diamond and Solomon (1959). A permeability barrier, a functional membrane, covered the outer surface of the cytoplasm of both *Nitella* and *Nitellopsis*. Electron microscopy by now was producing pictures of the plasmalemma by the use of permanganate fixation: the 'unit membrane' covered plant as well as animal cells.

This has brought our historical survey quite close enough to the present day. Adopting the dogma that membranes generally control diffusion rates and produce observed potential differences and resistances, we can proceed to look at the work of the past two decades.

CHAPTER 3

Water relations

Introduction

The study of the water relations of plant cells has suffered from many difficulties of experimentation and interpretation. These difficulties are caused by the complexities of the paths of flow of water and solutes in multicellular tissues. As well, water flow at the cellular level has usually been associated with major changes in cell volume and hence, probably, of water permeability, during osmotic shrinking and swelling.

In studies with giant algal cells, these troubles have been circumvented in two ways, both of which depend on the size, shape and robustness of the cells. Osterhout (1949) exploited the long, cylindrical shape of cells of the charophyta when he developed the technique of transcellular osmosis, later made quantitative by Kamiya & Tazawa (1956). In these experiments the two ends of the cell are in different solutions separated by a greased barrier. A difference in osmotic potential between these solutions causes a transcellular flow of water inwards at one end, along the vacuole (and cytoplasm) past the barrier and out at the other end. A second class of cells, namely large and tolerant coenocytes such as *Valonia* and *Halicystis*, survive perfusion of the vacuole and a near-zero hydrostatic pressure between the vacuole and medium. Then, imposed osmotic potential differences lead to a change in water flux, from which cellular water permeability may be estimated.

Ironically, some of the giant cells, so convenient for work on osmotic water permeation, have a grave disadvantage when it comes to the measurement of diffusive permeation, that is, the steady-state exchange of isotopes of water. The permeability of cell membranes to water can be so high that diffusion resistances in the unstirred layers on either side of the cytoplasmic layer may constitute a high proportion of the total resistance. Hence, large and possibly imprecise corrections may have to be made to calculate the diffusive permeability of the membrane system itself.

Our use of the words 'membrane system' raises the further question of interpreting the data on water permeability in terms of the properties of the bounding membranes and cytoplasm separately. In fact no attempt has been made to do this for water, but it is possible for nonelectrolytes as we shall see in chapter 4.

At the same time as interest developed in using giant algal cells for the study of water relations, biophysicists became aware of the relevance of the 'thermodynamics of the steady-state' for the discussion of fluxes of water, molecules and ions through membranes. Some of the theory for this, and a refreshingly new look at basic mechanisms by which water might move through cell membranes, has been given by Dainty (1963a). This splendid review is essential reading for all interested in the subject matter of this chapter.

Osmotic potential of the vacuolar sap

Nearly all the vacuolar contents of giant algal cells are composed of chlorides of the common alkali metals; details appear in chapter 5. Even in marine cells, the osmotic potential (Π_1) of the vacuolar sap exceeds that of sea water by a few hundred kilopascals, the water potential being equalised by the hydrostatic pressure of the sap maintained by the mechanical strength of the cell walls. The presence of this turgor pressure (ΔP) is sometimes blatantly apparent when handling charophyte cells, as more often than might be expected by chance a drop of sap spurts into the observer's eye when a cell is cut or bent and broken. Turgor, rather than mechanical stiffening of the cell wall, is responsible for the erect posture of these plants under water.

The osmotic potential can be estimated directly from detailed chemical analyses of the sap, or from its freezing point depression. Indirectly, Π_1 may be got from measurements of turgor pressure and use of the equation:

$$\Psi_w = P - \Pi \tag{3.1}$$

where Ψ_w is the water potential, a measure of net escaping tendency of water in a phase (see appendix A). For non-growing charophyte cells in the usual dilute media, there is osmotic equilibrium between inside and out, hence $\Delta \Psi_w = 0$

TABLE 3.1 *Values for internal osmotic potential and turgor pressure, in some giant algal cells*

Cell	Π_i/MPa	ΔP/MPa	Medium	Method	Reference
N. flexilis	(0.77)	0.77	KCl 1 mM	Wedge deflection, plasmolysis	Tazawa (1957)
N. flexilis	0.62	(0.62)	–	$[K^+]$, $[Na^+]$, $[Cl^-]$ measured for sap	Tazawa & Nagai (1966)
C. corallina	(0.6)	0.6	APW	Wedge deflection	Barry (1967, 1970)
N. axillaris	(0.5)	0.5	–	Micromanometer	Green & Stanton (1967)
Valonia utricularis	(2.3)	0.12	SW	Pressure transducer	Zimmermann & Steudle (1970)
Chaetomorpha linum	5.3	(3.1)	SW	Cryoscopic (Depression of FP)	Zimmermann & Steudle (1971)
Chaetomorpha linum	4.6	(2.2)	SW	Cryoscopic	Steudle & Zimmermann (1971a)

Values in brackets inferred from osmotic equilibrium.

36

and $\Delta P = \Delta \Pi = \Pi_1$. ΔP has been measured in such cells by two methods: from the extent of elastic deformation of the cell under an applied weight, and from the volume of an air bubble in a micromanometer inserted into the vacuole. A few values of Π_1 and ΔP are collected in table 3.1.

Transcellular osmosis

When a difference in osmotic potential ($\Delta \Pi_0$) is established between the two external solutions in the transcellular osmosis apparatus, a complicated series of processes begins. Fortunately, as the theory shows (Dainty & Ginzburg, 1964a), between about 1 and 30 sec, depending on cell dimensions and $\Delta \Pi_0$, the rate of transcellular flow (dV/dt) is linearly proportional to $\Delta \Pi_0$ and to a coefficient of the cell, the hydraulic conductivity L_P. Thus:

$$dV/dt = L_P \Delta \Pi_0 A_{ex} A_{en}/(A_{ex} + A_{en}) = \tfrac{1}{2} L_P \Delta \Pi_0 A \qquad (3.2)$$

if A is the symmetrical area exposed at each end, instead of A_{ex} at the exosmosis end and A_{en} at the endosmosis end, in the asymmetrical case. L_P in m s^{-1}Pa^{-1} may be converted to an osmotic permeability coefficient with the conventional units m s^{-1} by:

$$P_{os} = L_P RT/\overline{V}_w \qquad (3.3)$$

where $\overline{V}_w$ is the partial molar volume of water.

Figure 3.1 shows the volume flow with time in a typical experiment. Transcellular osmosis causes no microscopically visible damage or change to the cell, although the appearance of chloroplasts may change after flow induced by a large $\Delta \Pi_0$, say 1.0–1.2 MPa. Otherwise, streaming is normal and cells survive for long periods. If the transcellular set-up is made rather asymmetrical, as shown below (fig. 3.2) the rates of flow in the two different directions may be different. This is now agreed to be due to a real difference in permeability of the membrane system according to whether there is exosmosis or endosmosis. It is presumably the result of some modification to the membrane such as dehydration, caused by the osmoticum. A full account of the apparent rectification of flow by a *Nitella* cell preparation is given by Tazawa & Kamiya (1966). That the osmotic potential of the solution can affect the value of L_P

was shown by Dainty & Ginzburg (1964a) by using the two arrangements shown in fig. 3.3 in turn, sucrose being an effective solute for the purpose. In the right-hand arrangement a smaller flow was observed, i.e. L_p fell as the osmotic potential was increased.

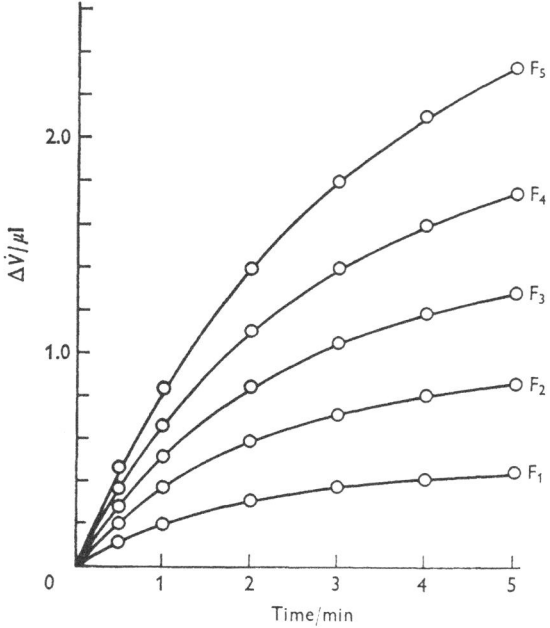

Fig. 3.1 Time course of transcellular osmosis in a single internodal cell of *Nitella flexilis*. F_1–F_5 refer to experiments when 0.1, 0.2, 0.3, 0.4 and 0.5 M sucrose solution was at one end and water at the other (from Kamiya & Tazawa, 1956).

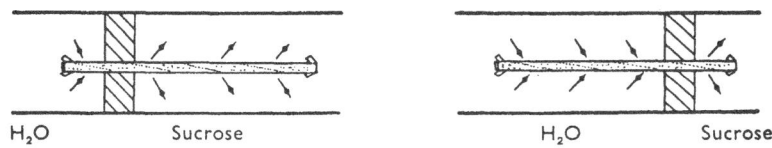

Fig. 3.2 The transcellular osmosis experiment, with unequal areas exposed to sucrose solution and distilled water.

When we come to examine actual values found for L_p in the charophyte genera *Nitella* and *Chara*, even under very different conditions an unusually close agreement is apparent. L_p is close to 1.0×10^{-12} m s^{-1} Pa^{-1}, which means that a flow of

100 μl per second, through 1 m^2 of surface, will occur in response to 0.1 MPa (1 atm) difference in Π_o. This is a measure of the net permeability of the cell wall, the two membranes and the cytoplasm in series. Using isolated cell walls and hydrostatic pressure gradients, the hydraulic conductivity of the wall, L_{Pw} has been found to be 2–4 pm s^{-1} Pa^{-1}. The hydraulic conductivity of the protoplast, L_{Pp}, is given by:

$$L_{Pp}^{-1} = L_P^{-1} - L_{Pw}^{-1} \qquad (3.4)$$

and its value is therefore approximately 1.5 pm s^{-1} Pa^{-1}.

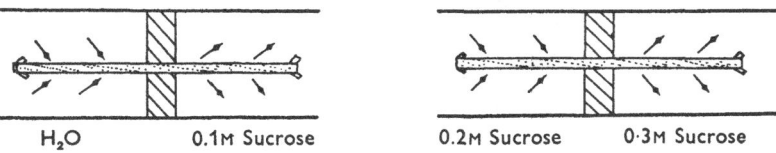

H$_2$O   0.1M Sucrose   0.2M Sucrose   0.3M Sucrose

Fig. 3.3   Illustrating experiments with transcellular osmosis in which appears an effect of the osmoticum on the water permeability of the cell.

Returning to fig. 3.1, it is seen that the initial flow rate decreases and eventually reaches a steady value after a few minutes. This decrease in rate corresponds to a gradual displacement of the solutes in the vacuole which travel with the transcellular flow, becoming in consequence diluted at the endosmosis end and concentrated at the exosmosis end. Such an effect is more magnified the longer and narrower the cell, and the steady-state rate may be low indeed, due to the virtual disappearance of water potential gradients across the membrane complex at each end. The process continues until diffusion just balances the opposite movement of solute carried by the transcellular flow. Tazawa & Nagai (1966) produced two 'cell halves' with very different vacuolar concentrations by tying off the central zone of a cell after a few minutes of transcellular osmosis. They then studied the ensuing changes in the vacuolar salt concentrations. Restoration of the lowered osmotic potential occurred over 5–6 days by means of light dependent uptake of K$^+$, Cl$^-$ and, to a lesser extent, Na$^+$; but this is more relevant to later chapters.

Osmotic flow during perfusion

The second method of determining L_P to be described is that used on *Valonia ventricosa* by Gutknecht (1967b, 1968b). The

39

general arrangement for perfusing *Valonia* is shown in fig. 3.4 and is the basic means of measuring fluxes of ions and molecules, osmotic and diffusive water flow, and electric p.d. and resistance of the cell.

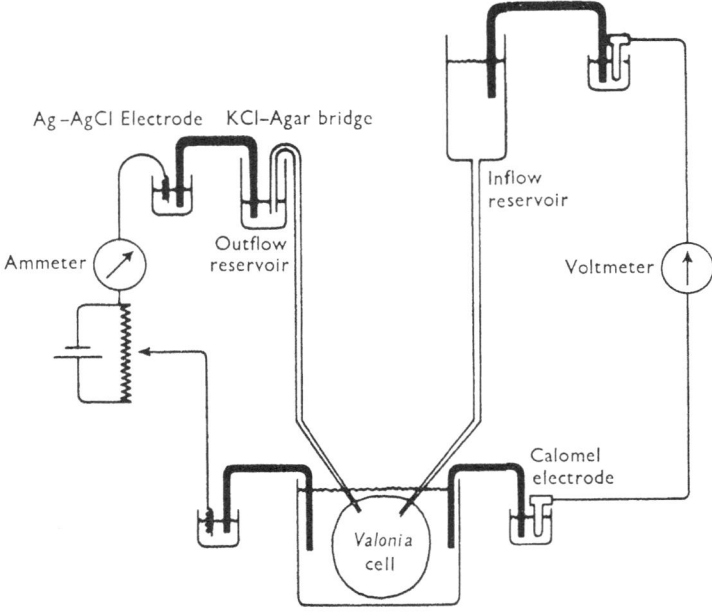

Fig. 3.4   Arrangement for vacuolar perfusion and electrical measurements in single cells of *Valonia* (from Gutknecht, 1967a).

In *V. ventricosa* there was found to be an 'endogenous', maintained volume flow into the vacuole, when it was perfused with an artificial sap isotonic with natural sap. Although in this instance $\Pi_1 - \Pi_0$ was about 0.1 MPa, the inward flow was not purely osmotic: the accompanying ion fluxes produced a virtually isotonic fluid (Gutknecht, 1967a) independently of the osmotic pressure difference. This is reminiscent of the phenomena in gall bladder secretion. By adding mannitol to the outside and noting the new volume flux, L_P was calculated from

$$L_P = \Delta J_v / (RT \Delta c) \qquad (3.5)$$

$RT \Delta c$ is the change in OP difference caused by the addition of mannitol. Readers are referred to Thain (1967) for a discussion of the Van't Hoff equation and osmosis generally. No correction was made for extra-protoplast resistance because, by analogy with the value of L_{P_w} for *Chara* and *Nitella*, it was assumed that the hydraulic conductivity of the wall of *Valonia* would be 500 times that of the whole cell of *Valonia*. This needs verification. The uncorrected L_P was 0.0183 ± 0.0027 pm s^{-1} Pa^{-1} (12 cells), and P_{os} was therefore 2.48 ± 0.36 μm s^{-1}. Note that this is much smaller than in charophyte cells.

Self-diffusion of water

If one measures the flux of isotopically labelled water (ϕ_w^*, in mol m^{-2} s^{-1}) across a membrane, this defines a diffusive permeability coefficient, P_d^*:

$$\phi_w^* = P_d^* \Delta c_w^* \qquad (3.6)$$

The usual assumption made is that the permeability coefficient for the labelled water, P_d^*, is substantially equal to that for normal water P_d. If water crosses the membrane only by independent diffusion, P_d is equal to P_{os}, previously defined as $L_p R T / \bar{V}_w$. If water can also cross the membrane in a way providing co-operative interaction – either by flow in channels or pores or by single-file diffusion in very narrow channels – the value of P_{os} will be greater than that of P_d. The problem has been treated from a statistical-kinetic viewpoint by Kedem & Katchalsky (1961) and by Lea (1963): it may be useful to consult Thain's textbook (1967) here. Comparisons of the two coefficients for animal cells have shown that P_{os} is greater than P_d, the ratio varying from about 1.2 to about 50 (Dick, 1966).

A fair comment is that P_d has rarely been measured adequately because of the ubiquitous unstirred layers, the unstirrability of which was not appreciated. The best evidence for a difference between P_{os} and P_d has come from work with red blood cells, which is embodied in a series of publications by Solomon and his group (Paganelli & Solomon, 1957; Sidel & Solomon, 1957; Rich, Sha'afi, Barton & Solomon, 1967; Sha'afi *et al.*, 1967). $P_{os} : P_d$ is about 2.4 in human red cells. The radius of equivalent cylindrical pores that would lead to such a ratio is 0.45 nm. Solomon (1968) may be consulted for a discussion of this and other methods of calculating an equivalent pore size. We refer again to the pore concept when discussing the permeation of nonelectrolytes in the next chapter.

In artificial, bimolecular membranes made of biological lipids, P_{os} and P_d appear to be equal (Cass & Finkelstein, 1967). When this is so, it may be concluded that narrow pores are absent and that water molecules permeate independently through the membrane by solution in the lipid.

Because of the importance of a comparison between P_{os} and P_d, in relation to the possible existence of membrane pores, Dainty & Hope (1959a) attempted to measure both permeabilities in C. *corallina* but succeeded only with the osmotic experiments. D_2O was used as a tracer for H_2O in self-diffusion studies, but the rate of exchange of the isotopic forms of water between vacuole and external medium was almost equal to that found using an agar rod of the same dimensions as the cell. Hence, clearly, the diffusion resistance of the unstirred layers of water was large compared with that of the membranes and the latter could not be estimated.

TABLE 3.2 *Some values of osmotic and diffusional permeability for water in various membranes*

Species	$P_d/\mu m\ s^{-1}$	$P_{os}/\mu m\ s^{-1}$	Reference
Chara corallina	–	200	Dainty & Hope (1959a)
Valonia ventricosa	2.4	2.5	Gutknecht (1967b)
Valonia utricularis	–	13	Steudle & Zimmermann (1971b)
Homo sapiens (red cell)	48	120	Solomon (1968)
Chaos chaos	0.23	0.37	Prescott & Zeuthen (1953)
Phospholipid bilayers	10	10	Cass & Finkelstein (1967)

It is possible that narrower charophyte cells, a well-stirred outside medium and the use of T_2O might make the experiment feasible: the weighing technique employed with D_2O–loaded cells required a relatively stationary medium. This experience makes the point that in any measurement of water permeability corrections must be made for the resistances of water layers adjoining the membrane. The time-honoured way of doing this is to measure the permeation of the substance in a killed cell, following measurements in the same, live cell. It is better to use a cell in which the bounding membranes have disappeared due to a natural physiological process. Such a process is the formation of aplanospores in *Valonia*, referred to in chapter 1.

Gutknecht (1967b) having measured the efflux of tritiated water from a 'normal cell', perfused with labelled, artificial sap, induced aplanospore formation by perfusion with sea water. Then the cell wall, internal and external unstirred layers of solution were identical in the two measurements while the plasmalemma and tonoplast were present only in the first. However, the cytoplasm layer had been replaced by a layer of small spheres each surrounded by a membrane, which may or may not be equivalent to the diffusion resistance of intact bulk cytoplasm. On balance, though, this procedure is preferable to killing the cell with heat or chloroform which may deposit lipids along the inside of the cell wall.

Using the preferred method, the diffusional permeability of the intact protoplast may be obtained from:

$$P_{dp}^{-1} = P_d^{-1} - P_{dw}^{-1} \qquad (3.7)$$

where p refers to the protoplast and w to the wall and un-stirred layers (Dainty, 1963a). Clearly the importance of the correction increases if P_{dp} is comparable with P_{dw}, which is presumably the case with *Chara*. However, in *Valonia*, P_d is much smaller and it turns out to be possible to measure it. Thus Gutknecht (1967b) found P_d was $1.22 \pm 0.03 \ \mu\mathrm{m \ s^{-1}}$, hence P_{dp} was $2.36 \pm 0.17 \ \mu\mathrm{m \ s^{-1}}$ (five cells). This is very nearly equal to P_{os} and there is therefore no reason to invoke aqueous pores as the chief means of water permeation in this genus. There is still the possibility of hydrophilic pores through which ions (and a small fraction of the water) might migrate. This will be taken up again in the next chapter and again in chapter 7 under electro-osmosis. The results for *Chara*, *Valonia* and some animal cells are compared in table 3.2.

CHAPTER 4

Permeability to nonelectrolytes

Introduction

As pointed out in chapter 2, it was data from charophyte cells and several other genera such as *Rhoeo* and *Beggiatoa* (a bacterium) that led to the notion that cell membranes present a diffusion barrier, effective according to the lipid insolubility of the solute. Evidence of the relatively high permeabilities of lipid-soluble nonelectrolytes contributed markedly to the Davson & Danielli model for cell membranes, of the 1930s. The only subsequent major study of molecular permeation in the Collander genre has been the monumental work of Diamond and Wright using gall bladder preparations (see Diamond & Wright, 1969; Wright & Diamond, 1969) and that of Tay & Findlay (1972) on toad duodenum.

Very nearly all the work that has been done on plant cells has treated the protoplasm as a single membrane separating the vacuole from the medium. The separate permeability coefficients of the plasmalemma and tonoplast have been found only for one or two solutes. As with water, all the early published 'membrane permeabilities' are for the protoplast. Now that it is possible to use nonelectrolytes labelled with ^{14}C it is to be hoped that an extensive study of the separate permeabilities of the two membranes will be undertaken following the method of Dainty & Ginzburg (1964b).

The work of the Finnish school

The permeability coefficients of numerous nonelectrolytes in *Chara ceratophylla* were measured by Collander & Bärlund (1933). Further measurements were made with certain smaller molecules (Collander, 1949a); Wartiovaara (1949) and Collander (1954) used *Nitella mucronata* to obtain more accurate results for some rapidly-permeating substances because the smaller diameter of *N. mucronata* shortened the diffusion path in the vacuole. Permeability coefficients were calculated from chemical determinations of the amount of a substance diffusing out of a previously loaded cell or, for very slow permeators, into a cell.

The care and ingenuity involved in such work (with up to 70 compounds) can only be imagined by those used to an electronic machine tapping out the result of a radioactive assay. In the case of certain rapidly-permeating compounds, a plasmometric method was used to determine permeability, and especially the rank order of permeability between certain amides.

Calculation of permeability coefficients

Permeability coefficients were calculated from concentrations assuming surface-controlled diffusion. For an efflux experiment this predicts a negative exponential approach to equality of concentrations inside and outside the cells:

$$c(t) = c(o)\exp(-PAt/V)$$

whence

$$P = (V/At) \ln[c(o)/c(t)] \qquad (4.1)$$

where $c(t)$ is the concentration of the cell sap after time t; $c(o)$ that at the beginning of the elution period; A, V cell area and volume; and P the permeability coefficient. Collander (1954) discussed the justification for this method, which assumes absence of active transport or changes in the molecule during permeation, and outlined the means of correcting for wall and unstirred layer resistances. This correction has already been referred to in chapter 3, where we pointed out the desirability of a better method than that employing heat-killed cells. A more precise correction is possible if the equations for 'permeation+diffusion' in a cylinder are applied in the way shown by Dainty & Ginzburg (1964d). This we return to, below. In the experiments described, the authors used the relation:

$$P^{-1}_{\text{protoplasm}} = P^{-1}_{\text{living cell}} - P^{-1}_{\text{dead cell}} \qquad (4.2)$$

to calculate $P_{\text{protoplasm}}$ from observations with living and subsequently-killed cells (see Wartiovaara, 1942).

45

Values so determined were compared with lipid:water partition coefficients, with various factors for molecular weight, in order to test the concept that the control of cellular permeability to nonelectrolytes was vested in a lipid layer (Overton, 1899). If this were the main factor, then permeability was expected to be proportional to the lipid:water partition coefficient. Such a general relationship was indeed found, as seen for *Nitella mucronata* in fig. 4.1. There are some notable

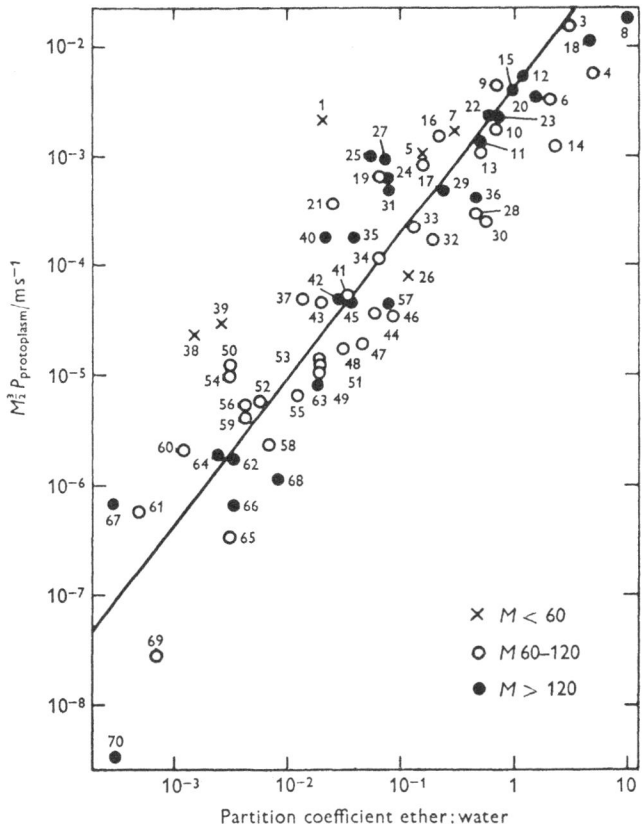

Fig. 4.1 The quantity $M^{3/2}P_{protoplasm}$ for seventy different molecules plotted against the partition coefficients between ether and water of the same molecules. M is the molecular weight of the molecule, and $P_{protoplasm}$ the permeability of cells of *Nitella mucronata* with a correction based on measurements with dead cells (from Collander, 1954).

deviations from the main pattern, which fall into at least two classes to be discussed below. Many of the partition coefficients had to be measured by Collander (1949b, 1950) because the data was unavailable elsewhere. In the absence of exact knowledge about the lipid composition of membranes, he used the non-polar solvents isobutanol, ether, olive oil and (olive oil +oleic acid).

An influence of the molecular weight of the permeating molecule on its permeability is expected from the kinetic theory of diffusion. We can write for a particular non-electrolyte:

$$P = KD/\delta$$

Where K is the partition coefficient membrane:water; D, the diffusion coefficient for the membrane, and δ its thickness. This is appropriate if diffusion within the membrane is rate-limiting, which, however, is not the only possible way of regarding the permeation process. Factors affecting D thus affect P. D is proportional to an average velocity attained during thermal agitation which in turn is proportional to $M^{-1/2}$ because for a single molecule:

$$M(\bar{v})^2/2\mathcal{N} \approx kT \qquad (4.3)$$

and therefore $\bar{v} \propto M^{-1/2}$.

kT is approximately the mean kinetic energy of translation and $\mathcal{N}$ is Avogadro's number. Other considerations lead to $P \propto M^{-1/3}$ for spherical molecules and to some greater power for non-spherical molecules. Measurements of diffusion coefficients for a range of large molecules in aqueous solution show these expectations to be valid. Clearly, however, as molecular weights of the molecules under consideration vary only by a factor of about 25, and K by a factor of more than 10^4, almost the same correlation is obtained between P and K, and between $PM^{1/2}$ and K. Collander (1954) suggested that the best fit was obtained by plotting $PM^{3/2}$ against K, for *Nitella*. In fact, when later results with very high molecular weights are included, least scatter is apparently obtained if the molecular weight term corresponding to equation 4.3 is $M^{-2.5}$ or M^{-3} (Diamond & Wright, 1969); the qualitative explanation of this is said to lie in a discrimination within the rather ordered membrane lipids against the compounds with higher molecular weight, which must cause more displacement of membrane components as they diffuse.

Deviations from the main permeability sequence

(i) Small molecules such as water, formamide, acetamide and dimethylformamide have higher permeabilities than expected from their partition coefficients, but the urea series is not anomalous, as it is in gall bladder (Wright & Diamond, 1969).

Wartiovaara & Collander (1960) argued that these 'deviant' molecules follow the same paths as the other solutes through the membrane lipids, but that a sieve effect restrains the larger nonelectrolytes. It is more likely, however, that these small molecules, which are all rather polar, do not penetrate the membrane by diffusion only among hydrocarbon chains of lipids, but through special hydrophilic regions of the membrane. Diamond & Wright (1969) discuss in more detail the arguments for this 'polar route' for small molecules. Whether these polar regions correspond to permanent or semi-permanent pores in the membrane is impossible to answer in the context of non-electrolyte permeability. However, when this and other suggestive evidence, for example electro-osmosis, is considered together, the case for hydrophilic pores in membranes of the charophyta may be rather stronger. It is worth noting that methanol, which is small and polar, is not off the main sequence in the correlation pattern of fig. 4.1 for *Nitella*, but is well above it in *Chara* (fig. 1 of Collander, 1949a).

(ii) Branched-chain members of a homologous series, such as tertiary-butanol or trimethyl citrate, tend to permeate more slowly than their oil:water partition coefficients would predict. An explanation of this in terms of extra steric hindrance (sieving) in a membrane containing not random but regularly-oriented lipid molecules, seems plausible. Thus the partition coefficient in the membrane lipids is reckoned to be less than in a bulk phase.

Temperature coefficients for nonelectrolyte permeation

The general observation is that, although the molecules are diffusing and not undergoing active transport or metabolically-mediated chemical changes to permeate *Chara* or *Nitella* cells, the Q_{10} for P is 2–5 in the temperature range 10–20 °C (Wartiovaara, 1942, 1949). The small, rather polar molecules, significantly, have smaller Q_{10}s. The temperature dependence of P of course contains the temperature dependences of both K and D and hence interpretation of such data is difficult.

K is related to the free energy change for the transfer of one mole of a particular solute from the aqueous to the lipid phase through the equation:

$$K = \exp(-\Delta G_{wl}/RT) \qquad (4.4)$$

(Diamond & Wright, 1969). It will be seen from (4.4) that $d \ln K/d(1/T) = -\Delta H/R$ and that the slope of an Arrhenius type of plot leads to a value for the enthalpy of activation and, if extrapolation is warranted, to an entropy of activation. Interpretation of these quantities depends on models and assumptions and further discussion does not seem rewarding at present (but see Zwolinski, Eyring & Reese, 1949). The same difficulties arise in the consideration of effects of temperature on ionic permeation, as seen in chapter 7.

The germinal nature of the work on nonelectrolytes with the giant algae is evident from reading the papers on gall bladder, already referred to. Also, the probability of widespread applicability of the permeation rules discussed above has been raised appreciably by the findings with gall bladder which are similar in all respects except that the urea homologous series is in the class of anomalous, quickly-permeating polar solutes. The duodenum of the Queensland toad (*Bufo marinus*) also has a permeability sequence similar to those of the other systems mentioned (Tay & Findlay, 1972).

The permeability of the plasmalemma and tonoplast to nonelectrolytes

Dainty & Ginzburg (1964*b*) used an elution procedure, and radioactive urea to get estimates of the permeabilities of the plasmalemma and tonoplast of *Nitella translucens*. The shape of the elution curve, of radioactivity remaining in the cell against time, can be used to calculate rate constants for washing out the urea from the cytoplasm and vacuole, if a number of assumptions are made:

(i) The cell behaves as two compartments connected in series, with diffusion within those compartments that is fast compared with diffusion across the rate-limiting boundaries to the cytoplasm.

(ii) Urea is not actively transported or modified by chemical reactions during its uptake and elution. There is reason to suppose that these assumptions are appropriate for many non-electrolytes, at least for the time-scales used in these experiments.

The mean values for P_t were 7.4 nm s^{-1}, and for P_p, 200 nm s^{-1}. The tonoplast is therefore rate-limiting for loss of urea over long periods. Approximate agreement between P_t and Collander's ($1949a$; 1954) values for permeability of *Nitella* and *Chara* protoplasts, which were about 2 nm s^{-1}, is satisfactory, considering the completely different techniques. Unless many further such studies are made, it will be difficult to further our ideas about differences in behaviour of the two membranes towards nonelectrolytes.

As with L_p (chapter 3), the osmoticum sucrose was found to decrease P_p by a substantial amount, by 24.5% (0.5 molal sucrose) to 67.5% (1.0 molal). Such effects were attributed to the swelling and shrinking of membrane proteins according to the water potential of the surroundings.

Permeability to alcohols, and corrections for extra-protoplast resistance

Dainty & Ginzburg ($1964c$) studied the permeability to several alcohols of *N. translucens* and *C. corallina* in connection with a consideration of the reflection coefficients of membranes of these cells, which we discuss more fully in the next section. Since the alcohols are rapidly permeating substances, correction for the resistance of the cell wall and other unstirred layers becomes quite crucial, as can be seen from Collander's data (1954), where for methanol $P_{\text{living cell}} = 3.23$ μm s^{-1}, $P_{\text{dead cell}} = 7.5$ μm s^{-1} whence $P_{\text{protoplast}} = 5.7$ μm s^{-1}.

The more precise corrections proposed by Dainty & Ginzburg are based on solving the equations for diffusion inside and outside the membrane system so that the proper driving force across the membrane can be found. $\Delta c_s(\text{memb.}) < \Delta c_s$, as fig. 4.2 shows. The underestimation of P inherent in using equation (4.1) arises, of course, from the fact that the driving force is actually $\Delta c_s(\text{memb.})$ and not Δc_s. It was shown, for diffusion in a plane (the cylindrical case proved intractable) that the ratio of the true permeability to that calculated from use of equation (4.1) is

$$\text{Ratio} = \theta / \int_0^{\theta^2} \exp \theta^2 \, \text{erfc} \, \theta \, d\theta^2 \qquad (4.5)$$

where $\theta^2 = 4P^2t/D$, erfc θ is an error function, and t the time of immersion in a radioactive alcohol during the influx experiments. A corrected permeability P' was obtained by a method of successive approximations with the aid of equation (4.5).

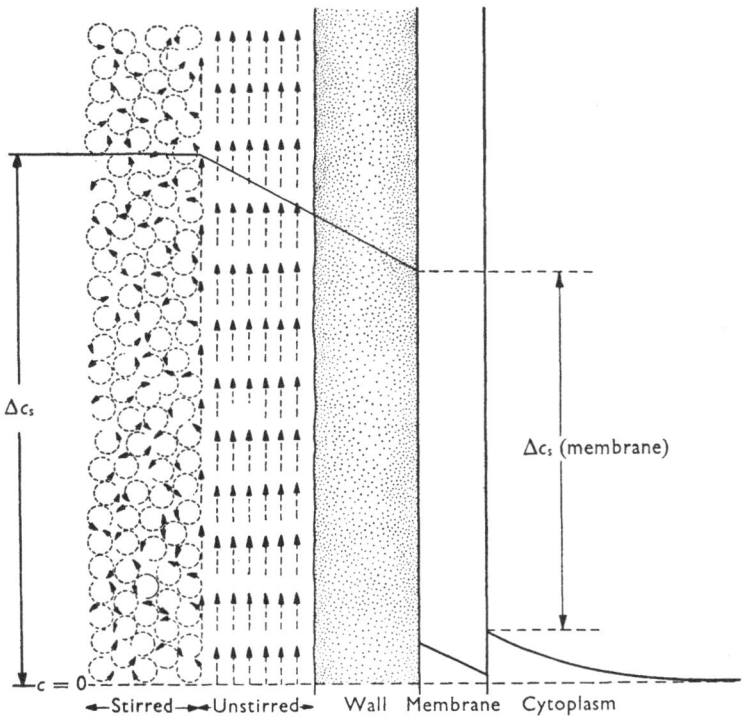

Fig. 4.2 Showing diagrammatically the concentration gradients of the solute s in the unstirred layers on each side of the plasmalemma in a plant cell. The effective concentration difference across the membrane is Δc_s (membrane), which is less than Δc_s.

It is interesting to compare the results of Dainty & Ginzburg (1964c) with those of Collander (1954) for *N. mucronata*; both the coefficients themselves and the magnitudes of corrections,

c

i.e. the ratios P'/P and $P_{\text{protoplast}}/P_{\text{cell}}$ respectively, are not too dissimilar (table 4.1).

TABLE 4.1   *Permeability coefficients for charophyte cells for several alcohols*

		$P/\mu\text{m s}^{-1}$			
		C. corallina	*N. translucens*	*N. mucronata*	
Methanol	P	2.94 ± 0.1	2.85 ± 0.1	P_{cell}	3.2
	P'	4.02 ± 0.4	4.82 ± 0.1	$P_{\text{protoplast}}$	5.7
Ethanol	P	1.97 ± 0.1	2.72 ± 0.1	P_{cell}	3.0
	P'	2.82 ± 0.3	4.25 ± 0.5	$P_{\text{protoplast}}$	5.5
Isopropanol	P	1.53 ± 0.1	1.72 ± 0.3	P_{cell}	2.3
	P'	2.04 ± 0.3	2.06 ± 0.5	$P_{\text{protoplast}}$	3.8

The reflection coefficient

As is well known, the effectiveness of a permeating solute in causing an osmotic volume flow across a membrane is lower than that of an impermeant solute. The relevant equation (see Thain, 1967) is:

$$\mathcal{J}_v = L_p(\Delta P - \sigma \, \Delta \Pi) \qquad (4.6)$$

where the coefficient σ is the reflection coefficient, whose value depends on the properties of both membrane and solute (see appendix A). It is a pure number, being 1.0 for impermeant solutes and less than 1.0 for permeant solutes. The decrease in volume flow represented by values of σ less than 1.0 is due to a volume flow of solute in the opposite direction to the water flow; and further, to any interaction of solute flow and water flow. It is possible to derive a general expression for σ in terms of membrane properties (Dainty & Ginzburg, 1963), and to use measurements of σ, P_{os} and P_{s} to test for the existence of solute–solvent interactions which might result from pores (Dainty & Ginzburg, 1964d).

Consider the model already suggested by the pattern of nonelectrolyte permeability, namely alternative hydrophobic (non-polar) and hydrophilic (polar) paths for the diffusion of various solutes including water which is not the least lipid-soluble substance. Then we can define the following quantities, some already encountered in chapter 3 or above:

K_{s}^{c} the partition coefficient for the solute, between the polar phase (c) of the membrane, and aqueous solution,

f_{sw}^c a frictional coefficient, representing the force on one mole of solute at unit relative velocity between the solute and water molecules in the c phase,

f_{sm}^c a frictional coefficient, defined as for f_{sw}^c, for interaction between solute molecules and the wall of the c phase,

P_s the permeability coefficient of the solute,

$\bar{V}_s$ the partial molar volume of the solute,

$\bar{V}_w$ the partial molar volume of water.

The frictional coefficients were first introduced for use in the flow equations in steady state thermodynamics by Spiegler (1958). In terms of the above quantities:

$$\sigma = 1 - (P_s \bar{V}_s / P_{os} \bar{V}_w) - K_s^c f_{sw}^c / (f_{sw}^c + f_{sm}^c) \qquad (4.7)$$

the second term being related to the contribution of the volume flow of the permeable solute to the total volume flow. This equation is appropriate even if the permeation of water and solute is not confined to the polar phase. Clearly f_{sw} is likely to be lower in the lipid phase than in c due to the more independent diffusion of solute and water, unless the pores are quite enormous ones. The derivation of (4.7) was given by Dainty & Ginzburg (1963), and a more general model was discussed by Kedem & Katchalsky (1961).

We now come to the point of the exercise. If the membranes of *Nitella* and *Chara* contain a significant number of polar pores, in which water and small nonelectrolyte molecules can interact, then the third term on the right-hand side of (4.6) should be greater than zero and $\sigma < 1 - P_s \bar{V}_s / P_{os} \bar{V}_w$.

To test whether this was so, Dainty & Ginzburg (1964*d*) measured σ for several alcohols and esters, by noting the concentration of the permeating solute that resulted in zero volume flow in the transcellular osmosis apparatus (chapter 3) when 0.1 M sucrose was at one end and the test solute at the other. Once again, due to unstirred layer effects in the test solutions, the driving force quickly drops with time as the solute permeates, and σ will be under-estimated since the condition of zero volume flow cannot be verified for some seconds. Corrections for this were made from knowledge that Δc_s was the following function of time, under these conditions:

$$\Delta c_s = c_s \exp \theta^2 \operatorname{erfc} \theta \qquad (4.8)$$

(with $\Delta c_8 = c_8$ at $t = 0$) these functions having already been defined. Corrected values for σ were:

Methanol	Ethanol	Isopropanol
0.5	0.45	0.4

$P_8 \bar{V}_8 / P_{08} \bar{V}_w$ had a maximum value of 0.1–0.15 for these solutes, using the values for P_8 discussed above, and inserting the P_{08} appropriate for $N.$ *translucens* (chapter 3). Hence the calculated value of σ is greater than 0.85.

Thus the conclusion is reached that the frictional interaction term is appreciable, and water-filled pores may indeed be a feature of the membranes of *Nitella*. See Rich *et al.* (1967) for similar conclusions from σ measurements on red blood cells.

The analysis that we have summarised above was ingenious and it is a shame that one has to agree with the authors that the conclusion is not as firmly based as it might seem, because of the following considerations.

(i) The measurements related to two membranes in series, the individual properties (P_{08}, P_8, σ) of which were not accessible.

(ii) The corrections to P_8 and σ for unstirred layer effects were large and not entirely satisfactory.

(iii) We may note that the alcohols used were not particularly 'anomalous' molecules in $N.$ *mucronata* in the sense of permeating much more quickly than their partition coefficients would predict, even though they are small, polar molecules.

Thus in fig. 4.1 above, isopropanol is almost on the line of best fit, ethanol is only a little above it and methanol a little further above. The deviations are greater in Collander's figs. 1 and 3 (1954) where P is plotted rather than $PM^{1.5}$ as in fig. 4.1. In the studies of Collander & Bärlund (1933) and Collander (1949a) with *C. ceratophylla*, only methanol is shown as being much more permeable than a molecule in the main sequence with the same K. Water is in all instances 2–3 orders of magnitude up the permeability scale from the main sequence.

Some measurements have been made of reflection coefficients in other systems. Zimmermann & Steudle (1970) found for *Valonia utricularis* the following values of σ; the molecular radii were quoted by these authors and we have added some partition coefficients.

54

Substance	σ	r/nm	K(ether)	$KM^{-1/2}$
Raffinose	1.0	0.61		
Saccharose	1.0	0.53		
Glucose	0.95	0.44	4.5×10^{-6}	3.4×10^{-7}
Glycerol	0.81	0.274	6.6×10^{-4}	6.9×10^{-5}
Acetamide	0.79	0.227	2.5×10^{-3}	3.2×10^{-4}
Urea	0.76	0.203	4.7×10^{-4}	6.1×10^{-5}

When compared with the main sequence of σ against $KM^{-1/2}$ in gall bladder membranes (see fig. 4, Diamond & Wright, 1969) the last three substances are certainly 'anomalous' in that σ is lower than expected. The correlation with molecular radius suggested to Zimmermann & Steudle that a pore model might be appropriate. However, in view of Gutknecht's conclusion of absence of pores (at least in relation to water permeation) in another *Valonia* species, we should not come to hasty conclusions. Determination of σ in *Valonia* for a large number of substances is clearly needed.

CHAPTER 5

Compartments and ionic concentrations

Introduction

In their interactions with the environment, algal cells require a traffic across their membranes of mineral ions and gases. There seems little evidence that there is a major requirement for organic molecules to cross the surface. It further simplifies the problem, that all environmental ions are free in solution, unlike the situation with plant roots in the soil. Since the cells thrive with a supply of mineral ions, water, carbon dioxide and light energy for their entire existence, it is appropriate for us to spend a good deal of time discussing ions. It is natural to ask the following questions. How do ions get into and out of cells? In what concentration and in what physico-chemical state are they in cells, and in various parts of cells? It will be seen that only partial answers can be given to these questions.

The approach which is to be taken in this chapter is to outline in a general way the processes of importance that are involved when ions move from the external medium (pond water, seawater or one of the artificial versions of these) through the cell wall and the membranes and so on to the vacuole. These processes are studied in practice by adding a few traceable ions of the species in question to one phase in order to follow their movement into other phases of the system (chapter 6). Such measurements are coupled with measurements of ionic concentrations and/or activities in those phases that are accessible (this chapter). The electric potential differences and conductances between such phases are also important quantities (chapter 7).

The cell wall

From the present point of view, the important properties are the ionic diffusion coefficients in the wall, its ion-exchange properties and its spatial homogeneity. About the marine giant algal coenocytes there seems to be little definite information of this kind, but in *Chara* ions move into and out of cell walls as if through an inhomogeneous cation-exchange phase. The fixed anionic groups originate from the ionisation of carboxyl groups of pectinic acids (Dainty, Hope & Denby, 1960). Dainty & Hope (1959*b*) represented the inhomogeneity of cell walls in terms of a system of charged pores containing ions and water: large macropores and small micropores (fig. 5.1).

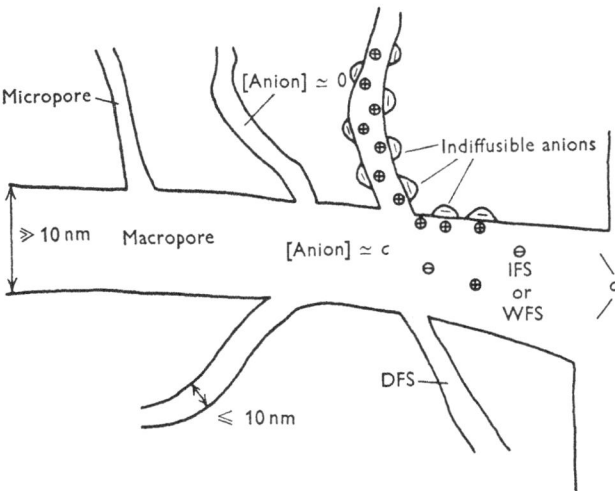

Fig. 5.1 Diagram of a proposed model for the cell wall of cells such as *Chara*, in terms of macropores and micropores, based on experiments on ion and non-electrolyte exchange between isolated cell walls and external media. IFS = Iodine free space; WFS = water free space; DFS = Donnan free space; *c* = external concentration (from Dainty & Hope, 1959*b*).

This idea comes from the dependence they found of the fraction of quickly-exchangeable calcium on the ionic strength. At low ionic strength, most of the calcium in the cell wall is only slowly exchangeable for univalent ions and it is supposed to be rather tightly bound to the walls of the macropore. This may involve the formation of a Stern layer (Davies & Rideal, 1961). At high ionic strength much of the wall calcium is in the quickly-exchangeable fraction. Calcium in the micropores is thought of as being always only slowly exchangeable.

The reluctance of ions such as Ca^{2+} to exchange, which may correspond to a low mobility in the wall, is recognised as the explanation of low apparent values for the exchange capacity of the wall. For example, Tyree (1968) arrived at the value 0.04 equiv (1 cell wall)$^{-1}$ from measurements of the resistivity of wall samples from $N.$ $flexilis$. Figures ten times as large as this come from isotopic exchange measurements on $Chara$ walls (Dainty & Hope, 1959b).

Whether this model of the Donnan region of the cell wall will usefully predict the rates of penetration of ions to the cell membrane from the outside solution is not known. Very little information is available on the diffusion of ions across cell walls, and such as there is, is generally in the form of isolated observations. Thus Mailman & Mullins (1966) estimated the diffusion coefficient for chloride in the cell wall of $Nitella$ $clavata$ as 5×10^{-12} m^2 s^{-1} with the cell bathed in 2 mM NaCl with 2 mM of divalent cations. This value came from the time-course of chloride concentration outside the cell immediately after an action potential. Since the value of the diffusion coefficient for chloride in water is about 2×10^{-9} m^{-2} s^{-1}, this represents a reduction of some 400-fold in the cell wall. The chloride concentration in the Donnan region of the wall may be some 1/10 to 1/100 of the value in the solution, which would yield apparent diffusion coefficients of 2×10^{-11} to 2×10^{-10} m^2 s^{-1}. Perhaps some steric factors are involved as well.

For cations, the rate of change of membrane potential in $C.$ $corallina$ when the ratio of [K$^+$] to [Na$^+$] in the medium is changed (Hope & Walker, 1961) suggests an apparent diffusion coefficient of 3×10^{-13} m^2 s^{-1} which is an average for the mutual exchange of K$^+$ for Na$^+$. Because of the Donnan effect of concentrating the counterions in the cell wall, its equilibrium is in fact slowed up by a factor equal to the Donnan ratio provided the whole process is rate-controlled by diffusion in an unstirred film (Dainty & Hope, 1959b). This ratio would have been several hundred (see table 5.1).

An estimate of the relative numbers of monovalent and divalent counterions in the wall when it is brought to equilibrium with various media can be got from the theory of the Donnan equilibrium (Briggs, Hope & Robertson, 1961). We list these in table 5.1 for some common media used in experiments with the charophytes. A more realistic theory, however, treats the ion-exchange system as inhomogeneous, with charged

TABLE 5.1 *Some experimental media used in work with charophyte cells, and the equilibrium concentrations expected within a homogeneous cell wall Donnan phase with the maximum concentration of indiffusible anions* (α) *assumed to be* 500 *mequiv* (l *wall water*)$^{-1}$. *The potential* ψ_{wo} *is calculated from the Donnan distribution ratio*

	External medium				Cell wall				
	[ ]/mM Ca^{2+} +				[ ]/mmol(l wall water)$^{-1}$ Ca^{2+} +				
	K$^+$	Na$^+$	Mg^{2+}	pH	K$^+$	Na$^+$	Mg^{2+}	pH	ψ_{wo}/mV
APW ^{a}	0.1	1.0	0.1	5.5	4.7	47	222	3.8	−97
APW ^{b}	0.1	1.0	0.25	5.5	3.0	30	225	4.0	−86
Hope & Walker (1961)	0.1	1.0	0	5.5	37	370	0	2.9	−149
FPW ^{c}	0.2	2.0	0.05	5.5	12	118	174	3.7	−113
Kitasato (1968)	0.1	2.0	2.0	5.5	1.1	22	238	4.5	−60
BAPW ^{d}	0.1	1.0	0.05	7.8	4.8	48	224	5.1	−97.5

^{a} Used by MacRobbie (1962) and others.

^{b} An artificial pond water to match approximately the field pond water for *C. corallina* (Hope & Walker, 1960).

^{c} Designed to increase the wall concentrations of monovalent cations (Walker & Hope, 1969).

^{d} Bicarbonate artificial pond water (Spanswick, 1970a).

We have taken into account, approximately, the changes in indiffusible anion concentration that occur due to adjustments in the equilibrium:

$$[A^-]+[H^+] \rightleftharpoons [HA] \qquad ([A^-]+[HA] = 500\ \mathrm{mequiv}\ l^{-1})$$

assuming pK_a to be 2.2 (Dainty, Hope & Denby, 1960). The changes are slight except for the solution of line 3. The cell wall potential is discussed further in chapter 7.

surfaces spaced apart by regions where the electric gradient is nearly zero (Overbeek, 1956). Dainty & Hope (1961) showed that more accurate predictions could be made about the exchange of sodium for calcium ions in cell wall segments, but many uncomfortable assumptions were necessary in using the theory. In summary of the properties of *Chara* cell walls we can list the following points:

(i) Uncharged substances, at least as large as mannitol, have access to most of the space in the cell wall: more than 95% of the wall water acts as 'free space' for mannitol.

(ii) Anions are largely confined to macropores, with small concentrations, of the order of 10 μequiv l^{-1}, in the equivalent Donnan space. For example, iodide has ready access to about 45% of the wall water, of which not much is in a surface film.

(iii) Cations, as well as having access to the anion free space in equal concentration to the mobile anions there, are counter-ions to the fixed anions. The wall is a reservoir of cations of

59

amount c. 0.01 equiv m^{-2} of surface area; most of these cations are calcium or magnesium when the growth medium or experimental medium has divalent ions in it (table 5.1). In the marine red alga *Porphyra*, by way of comparison, the amount of cation (mainly sodium) available for exchange between cell walls and sea water is about 40 mequiv (kg FW of killed cells)$^{-1}$. Giant cells of marine algae may be similar but the cell wall material is rather varied. Exchange of wall cations with those in seawater is very much faster than in the charophyte cells.

(iv) Because cell walls are porous, ion-exchange systems, they exhibit electrokinetic phenomena, which must be considered when dealing with such effects in intact cells (chapter 7).

The plasmalemma

In chapter 2 we referred to the wavering amongst plant physiologists in the 1950s, as to whether the plasmalemma was a physiological membrane in the sense of having an ion-restraining function. While there may now be agreement about the plant cell plasmalemma, a few workers with animal cells, notably Ling (see, for example, 1965), claim that much of the evidence about the movement of ions in such cells is interpretable as bulk-protoplasm-limited diffusion. In view of the known behaviour of counterions in ion exchange systems, where the activity coefficient may be low (Mackie & Meares, 1955) it would be unwise to ignore the possibility that ions have a diffusion coefficient in cytoplasm much less than in free solution. On the other hand there seems little doubt that the interfaces between plant cell cytoplasm and the external medium and the vacuole have a very high resistance to ion diffusion. The evidence for this has come largely from giant algal cells and is also discussed in a later chapter. We have already encountered the evidence about the diffusion of nonelectrolytes that also must be interpreted in this way.

The tonoplast

The electrical evidence implicating the tonoplast as a separate high-resistance membrane is described in chapter 7. The relative inaccessibility of this membrane and the difficulty with which changes in ionic composition can be made in the phases on either side of it have meant that less information is available about it than about the plasmalemma. We have already seen that the permeabilities of plasmalemma and tonoplast to urea have been calculated: this and the values of electric potential difference and resistance are all we know as yet of the properties of this membrane.

Studies of the ionic content of cells

Vacuolar sap

Giant algal cells were recognised a long time ago as possible sources of information about the accumulation of ions. Several millilitres of sap can be obtained from the vacuoles of some of the marine coenocytes (*Valonia, Halicystis*) and enough microlitres for analysis with modern techniques can be got from many other single cells. Sampling of a small fraction of one microlitre of sap using a dry glass microprobe was used by Vorobiev, Koltunov, Kurella & Li (1965) and others. Repeated sampling of this sort, of a live cell, may be possible. The sap so collected may be diluted for flame photometry, or a silver–silver chloride electrode inserted into it for determination of chloride activity.

TABLE 5.2 *Ion concentrations in the sap of mature coenocytes of* N. clavata

	K$^+$	Na$^+$	Ca^{2+}	Mg^{2+}	[ ]/mM Cl$^-$	SO$_4^{2-}$	PO$_4^{3-}$	NO$_3^-$	pH
				(From Hoagland & Davis, 1923)					
Pond water	–	0.2	0.8	1.7	0.9	0.35	0.004	0.5	5–9
Vacuole	54	10	10	28	91	8	3.7	–	5.0–5.2
				(From Hoagland & Davis, 1929)					
Pond water	0.51	1.2	0.65	1.5	1.0	0.33	0.0008		
Vacuole	43–59	40–86	5–9	5–11	101–107	6–10	2–4		

TABLE 5.3 *Ion concentrations in the sap of certain marine algal coenocytes* (from Blinks, 1929; Blinks & Jacques, 1929)

	Cl$^-$	Na$^+$	[ ]/mM K$^+$	Ca^{2+}	Mg^{2+}
V. ventricosa	600	34	568	–	–
V. macrophysa	600	92	506	1	–
H. osterhoutii	603	557	6.4	8.0	16.7

H. ovalis resembles *Valonia* in value of [K$^+$]/[Na$^+$], rather than *H. osterhoutii*.
Cation concentrations in *Valonia* were originally given as per cent of the halide present. We have assumed [Cl$^-$] is 600 mM (table 5.4). Sulphate was not detected.

TABLE 5.4 *Ion concentrations in the medium and in vacuolar sap, and electric potentials of vacuole and cytoplasm in some giant algal cells*

	$[K^+]_o$/mM	$[Na^+]_o$/mM	$[Cl^-]_o$/mM	$[K^+]_v$/mM	$[Na^+]_v$/mM	$[Cl^-]_v$/mM	ψ_{vo}/mV	ψ_{co}/mV	ψ_{vo}/mV	Remarks	Reference
Freshwater algae											
C. globularis	0.046	0.15	0.04	65±1.9	66±1.9	112±1.5	−181	–	–		Gaffey & Mullins (1958)
C. corallina	0.06	2.2	2.4	86±3 (8)	49±1 (8)	–	–	–	–	[Ca²⁺]₀ 0.16, [Ca²⁺]ᵥ 2.6. Cells stored in	
				74±2 (10)	47±2 (10)	106±12 (10)	−157±1.5 (10)	–	–	pond water	Hope & Walker (1960)
	0.1	1.0	1.6	64±4.5 (10)	57±3 (10)	–	−161±3 (10)	–	–	Cells stored in APW	
N. translucens	0.1	1.0	1.3	68±2.5 (10)	50±4 (10)	151±2 (15)	–	–	–	APW	MacRobbie (1962)
				78±1.5 (76)	60±1.5 (76)		−140±1.5 (12)	–	–	Cells from same pond as entry above	[Dainty *et al.*, pers. comm. to MacRobbie (1962)]
N. clavata	0.1	3.0	5.1	77±3	33±4	113±2	−122±2 (8–12)	–	–	Lower of two light intensities in culture, 6 cells for analyses	Barr & Broyer (1964)
N. flexilis	0.052	0.28	0.03	80±1.7 (43)	27.5±0.7 (43)	135±2.3 (16)	–	–	–	Culture medium = dechlorinated tap water (see table 5.5 for cytoplasmic layer and chloroplasts)	Kishimoto & Tazawa (1965a)
Hydrodictyon africanum	0.1	1.0	1.3	40±2	17±1.5	38±2	−90±4.5	−116±3	–	APW; cytoplasm ion concentrations also measured (table 5.5)	Raven (1967)
Tolypella intricata	0.45	1.0	–	90–110	3–10	120	−140 to −120	–	–	[Cl⁻]ᵥ attributed to Larkum (personal communication)	Smith (1968a)
	0.40	1.0	1.6					–	–		(Smith, Raven & Spanswick, unpublished data)

Brackish-water algae										
Nitellopsis obtusa	0.65	30	35	113±2 (69)	54±16 (63)	206±3 (45)	−120 to −200 (−122)	−	−	Artificial brackish water	MacRobbie & Dainty (1958)
Nitellopsis obtusa	1.0	24.5	38	95±6 (5)	60±4 (5)	174±4 (5)	−	−141±6 (18)	19±3 (18)	Artificial Hickling water	Findlay (1970)
Lamprothamnium succinctum	7.0	301	353	164±13 (7)	368±31 (7)	−	−	−	−	Natural brackish water	Kishimoto & Tazawa (1965b)
Lamprothamnium succinctum	7.1	288.5	337	252±15 (4)	136±14 (4)	372±27 (4)	−100	−	−	Artificial brackish water; cells cultured in this. (Mean conc. in cytoplasm layer also measured – table 5.5)	Kishimoto & Tazawa (1965b)
Marine algae										
Valonia ventricosa	12±0.2 (10)	508±5 (13)	596±2 (9)	625±5 (18)	44±1 (18)	643±2 (10)	+17±1 (16)	−70	(+87)	Cyt layer also analysed see table 5.5. ψ_{CO} measured from aplanospores or very young cells	Gutknecht (1966)
Chaetomorpha darwinii	13	500	523	540	25	(575)	+10	−32	−	Chloride on whole cells	Dodd, Pitman & West (1966)
Chaetomorpha darwinii	10	490	573	523±3 (10)	27±2 (10)	583±24 (10)	+5 or −29	−72	+77 or +43	ASW	Findlay et al. (1971)
Griffithsia pulvinata	10	490	573	535±15 (6)	30±4 (6)	606±19 (4)	−52	−86	+34	ASW, SEM for p.d.s <±3 mV	Findlay, Hope & Williams (1969)
Acetabularia mediterranea	10	470	550	355±20 (19)	65±2.5 (9)	480±30 (22)	−174	−	∼0	ASW	Saddler (1970)
Valoniopsis sp.	10	490	573	58±6 (10)	623±29 (10)	656±23 (10)	−5	−	−	ASW	Findlay et al. unpublished data

63

TABLE 5.5 *Ion concentration and chloride content of the cytoplasm and chloroplasts of giant algal cells*

	$[K^+]$/mM	$[Na^+]$/mM	$[Cl^-]$/mM	Q_{Cl}/mmol m^{-2}	Method	Remarks	Reference
Chloroplasts							
N. translucens	340	120	–	–	Vacuole 'blown out' of cut cell on basis of observed 4 μm layer		MacRobbie (1962)
N. flexilis	110±2.5 (26)	– 26±2 (26)	(240) 136±6 (25)	2.15 –	Vacuole 'blown out' of cut cell Removed vacuole and flowing cytoplasm by perfusion	% vol. of cell in chloroplasts and cytoplasm calculated from visual observations	MacRobbie (1964) Kishimoto & Tazawa (1965a)
C. corallina	–	–	(400)	0.90	Removed vacuole and flowing cytoplasm by perfusion	% vol. of cell in chloroplasts and cytoplasm calculated from visual observations	Coster & Hope (1968)
T. intricata	340	36	340	–	Freeze-dried and layered in non-aqueous density gradient		Larkum (1968)
Cytoplasm							
N. flexilis	125 (93–132)	9(0–16)	20(0–38)	–	Analyses after 'fast' and 'slow' perfusion, and analyses of centrifuged cytoplasm following nitrate perfusion		Kishimoto & Tazawa (1965a)

Species					Method		Reference
N. flexilis	—		14	—	Ag–AgCl microelectrode	Activity measured	Lefebvre & Gillet (1971)
C. corallina	115±10 (9)		10	—	Ag–AgCl microelectrode	Activity measured	Coster (1966)
Griffithsia pulvinata	153±6 (6)		—	—	K-responsive microelectrode	Activity measured	Vorobiev (1967)
			—	—	K-responsive microelectrode	Activity measured	Vorobiev (1967)
N. translucens	120	54	—	—	Centrifugation, flame photometry		MacRobbie (1962)
			65	—	Centrifugation, electrometric titration		Spanswick & Williams (1964)
			90	—	Centrifugation, Ag–AgCl electrode		Hope, Simpson & Walker (1966)
T. intricata	87–97	4–22	23–31	1.2	Centrifugation, flame photometry		Larkum (1968)
Cytoplasm (including chloroplasts)							
Hydrodictyon africanum	93±12	51±8	58±6	—	?	On basis of observed 20 μm layer	Raven (1967)
Lamprothamnium succinctum	137±7 (915)	46±10 (15)	86±19 (13)	—	Vac. sap removed by perfusion		Kishimoto & Tazawa (1965b)
Valonia ventricosa	434±9 (8)	40±2 (8)	138±4 (8)	—	Cells cut open, sap drained, remainder washed with isotonic buffer		Gutknecht (1966)
Acetabularia mediterranea	400±30 (6)	57±5 (19)	—	—	Flame photometry K+, isotopic equil. Na+		Saddler (1970)

The data of Hoagland & Davis on *Nitella* (table 5.2) and those of Blinks & Jacques on certain marine algae (table 5.3) were collected in the 1920s. These were the result of chemical analyses for the ions concerned. Later results (table 5.4) were obtained by flame photometric techniques or potentiometric or coulometric titration in the case of chloride. While the charophyte cells are not directly comparable with each other because of widely different culture media or treatments, the results for marine algae should be rather more constant because of the relative constancy of composition of sea water. In these results it is assumed that pure vacuolar sap has been obtained. The probability of contamination of sap with cytoplasm is of course much less than the reverse. Speed of operation is the essence of success since once a cell is cut exchange of ions between vacuole and cytoplasm may be accelerated. This may be peculiarly true of charophyte cells, in which it is known that the tonoplast is capable of conducting an action potential. Cutting the cell produces action potentials (MacRobbie & Walker, unpublished) and this may exchange ions rapidly between cytoplasm and vacuole.

Several general conclusions are possible from this data.

(i) Several species of ions in vacuolar sap, notably K^+ and Cl^-, sometimes Na^+, are in a concentration in excess of that in the medium outside.

(ii) Mature cells have a relatively constant ion content during their existence. No doubt this is partly due to their very large volume:surface area ratio.

(iii) Most, but not all genera have higher potassium concentration than sodium, and the discrimination in favour of potassium, expressed by $([K^+]_v/[Na^+]_v)/([K^+]_o/[Na^+]_o)$ is much greater than unity.

Exceptions to this rule occur in several genera (*Halicystis osterhoutii* and *Valoniopsis* sp.) where the ratio $[K^+]_v/[Na^+]_v$ is almost reversed, but no gradations between these two extremes have been recorded.

The cytoplasm

The cytoplasm, in mature cells of the types we are discussing, is a thin layer of great structural complexity. In chapter 1 the main features visible in electron micrographs were described.

Complexity in relation to ions may arise because of the following factors:

(i) Organelles are capable of active transport (Robertson, 1968).

(ii) The concentrations of diffusible and indiffusible ions are probably different in the various organelles and in the 'ground cytoplasm'.

(iii) Even after we have considered the more obvious organelles such as chloroplasts and mitochondria, the cytoplasm is still not homogeneous.

There is considerable difficulty, for example, in the concept of 'ion concentration in the cytoplasm' because of the presence of complex plane and tubular membrane systems. The soluble part of the cytoplasm is considered to contain indiffusible charges, as a result of its proteins having charged amino-acid constituent side chains, such as those of glutamic acid, aspartic acid, asparagine and glutamine. What indications do we have of the concentration of the indiffusible ions? It can be roughly estimated as the charge concentration needed for electric neutrality when all the concentrations of inorganic ions are summed. On this basis the answer is about 100 mM for $N.$ $flexilis$ (table 5.5) and nearly this for $Hydrodictyon$ and $N.$ $translucens$. Problems arise in obtaining such data: even in an enormous coenocyte of $Valonia$, a layer of cytoplasm 10 μm thick does not constitute much more than about 12 μl in a cell 20 mm in diameter. Gutknecht (1966) used the method of scraping the cytoplasm from inside a cell following cutting and flushing out the vacuole. Others have attempted to concentrate the cytoplasm by centrifugation (MacRobbie, 1962) or to remove the vacuolar sap by perfusing a sucrose or nitrate solution carefully from one end of a cut cell to the other (Kishimoto & Tazawa, 1965a). In $vivo$ measurements of ionic activity, using special electrodes specifically permeable to cations, have not been very productive. Use of Ag–AgCl microelectrodes has been more successful and the chloride activity of the flowing cytoplasm is now known within certain limits.

The distribution of ions between various phases of the cytoplasm was examined by Kishimoto & Tazawa (1965a). To do this they relied on the fact that in charophyte cells the mobile (streaming) cytoplasm can be perfused away by a relatively rapid perfusion stream, leaving chloroplasts still attached to the inside of the cell wall. The endoplasm or gelled layer of cytoplasm also remained (see chapter 1). Centrifugation at about 2000 g, as well as forcing the mobile cytoplasm to one end of the cell, strips the chloroplasts from the endoplasm and layers them below the clear cytoplasm. This causes an irreversible redistribution of the cell contents, unlike centrifugation at more moderate forces, when the clear cytoplasm will often resume streaming after being layered at one end of the cell (chapter 1). The most reliable of these estimates have been assembled into table 5.5. A different approach by the same authors was to perfuse the cell vacuole slowly with a solution of potassium and calcium nitrates, to tie off the perfused cell and to centrifuge it gently (200 g, 10 min) so that the flowing cytoplasm collected at one end. Enough is obtained for analysis, and the values for sodium and chloride concentration are low and reasonably repeatable. Contamination of the cytoplasm sample by sap containing large amounts of chloride and sodium is avoided in this way; but the loss of these ions from the cytoplasm during perfusion cannot be estimated. From table 5.5 it can be seen that:

(i) $[K^+]$ exceeds $[Na^+]$ in the cytoplasm, whenever it has been compared. Unfortunately there seem to be no estimates for cells of *Valoniopsis* or for *Halicystis osterhoutii*, the species in which vacuolar sodium exceeds potassium.

(ii) Chloroplast chloride exceeds in concentration that in the rest of the cytoplasm.

The central problem, which has always interested plant physiologists, is to find how these distributions arise and how they are maintained. Are ions accumulated while cells are young and merely locked in behind impermeable barriers during the adult life of the cells? Clearly this will not suffice, since rather similar concentrations are found in *Nitella* and *Chara* cells during elongation from a few millimetres to many centimetres length (MacRobbie, 1962). A clear picture of what is happening requires knowledge about the movement of ions across the controlling surfaces under various conditions. Thus we need to know as much as possible about the properties of the plasmalemma and tonoplast in relation to ions. Studies of the electrical properties of the membranes, as well as being of intrinsic interest, have contributed markedly to the admittedly incomplete picture of ionic relations in cells.

CHAPTER 6

Ionic fluxes and kinetic models

Introduction

The ionic concentrations discussed in the previous chapter arise by the diffusion in aqueous compartments, and by the transport across membrane barriers, of individual ions or of packaged groups of ions. The measurement of the rates at which these processes occur should in the end provide an explanation of the observed concentrations and indications of the mechanisms at work in the membranes. These goals, as we shall see, still lie ahead; their direction, at least, is known.

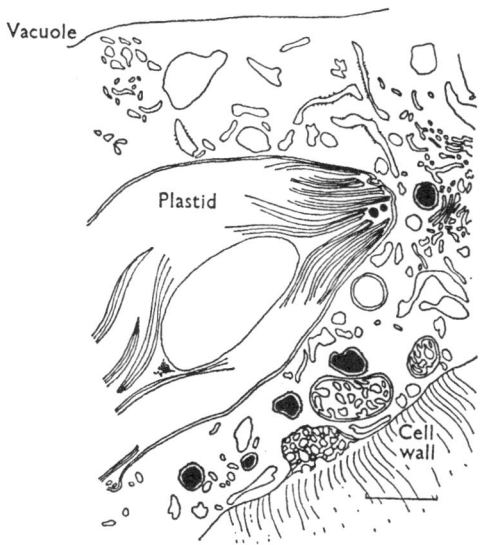

Fig. 6.1   Section of the cytoplasm of a cell of *Chara corallina*, from an electron micrograph of a glutaraldehyde-fixed cell.
Scale mark is equivalent to 1μm in length.

Most of the ions in question have radioactive isotopes which are, to the accuracy required here, chemically indistinguishable from the naturally-occurring forms. The radioactive isotope becomes a tracer when added suddenly to one of the compartments of a system: it allows the measurement of the flux of ions which originate in the labelled compartment and are found at a later time in another. The assumption of indistinguishability means that we can take the radio-active ions to be accompanied by inactive ones in the proportion in which they are found in the labelled phase (Y_a/Q_a).

The difficulties inherent in this approach are those of gaining access to the important phases in the system for labelling and for sampling: often the latter, when it is possible, involves the sacrifice of the cell.

Giant cells offer the vacuole as well as the outside medium for labelling and sampling, but no significant use has been made of the vacuole-labelling possibility to date. Nor has it yet proved possible to obtain serial samples from the same cell, except from slowly-perfused, marine algal cells such as *Halicystis* and *Valonia*. These methods, together with the measurement of the radioactivity of the whole cell and of the external solution, are the normal ones. They would be adequate for the measurement of the chief fluxes if it were not for the complexity of the compartments in the cytoplasm (fig. 6.1).

Two features of the cytoplasm indicated in this figure should be noted, although it is not yet possible to take full account of them. They are the large number of separate compartments, whose interconnections are not clear, and the variety of organelles which may well be involved in the transport of ions.

Plant cells and models

The plan of the plant cell provided by electron microscopy is disappointingly complex in compartments. Ion flux measurements are therefore interpreted in terms of the simplest model of compartment-arrangement that will work most of the time.

The simplest model, which was accepted as consistent with experimental results until about 1968, is shown in fig. 6.2. In this model the plasmalemma and tonoplast are rate-limiting, while free diffusion is assumed in the cytoplasm, wall and other aqueous media. The cell wall does not here formally appear in the model, being for this purpose regarded as an informal correction. This model yields equations which are analytically soluble only if the assumption is made that net fluxes are zero.

This generally restricts flux measurements to such steady-state conditions.

If enough detail could be obtained in the curve of radio-activity absorbed by the cell as a function of time, that taken up by the cell wall being allowed for, then the equations for the model would allow ϕ_p, ϕ_t, Q_c and Q_v to be found. More practically, one might allow the cell to take up radioactivity for a known time, and then elute it with a non-radioactive solution. From the efflux as a function of time, as Pallaghy & Scott (1969) have shown, one can obtain ϕ_p, ϕ_t, Q_c and Q_v (fig. 6.3, and see Hope, 1971). This approach was used for *Griffithsia* by Findlay, Hope & Williams (1969). However, most experimenters have measured Q_c and Q_v directly and used the predictions of the model to get ϕ_p and ϕ_t.

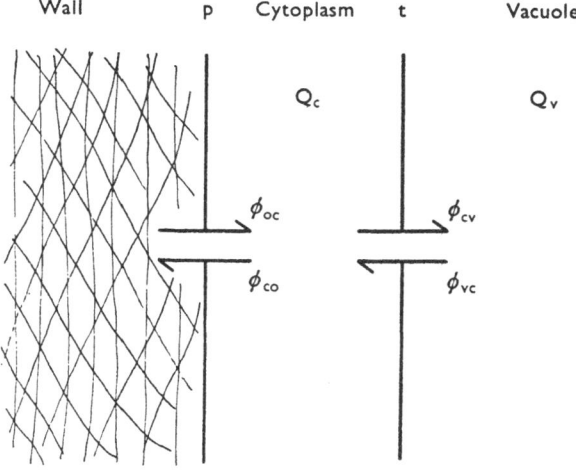

Fig. 6.2 The arrangement of compartments in series used as a model of a cell to describe the uptake of a substance after supplying it to the medium, or the loss of the substance to the medium from cells: p, plasmalemma; t, tonoplast.

For the initial period of uptake, while S_c is very much less than S_0, the rate of increase of cellular radioactivity will be given by $\phi_p S_0$. This method of measuring ϕ_p is so simple that it has been often used, although it is easy to over-estimate the 'safe' uptake time and hence to under-estimate the influx.

The model predicts that the radioactivity in the vacuole, Y_v, will rise in a curve approximating a square-law, followed by a linear increase. This time-course is often approximated by a linear rise after an initial lag. The results of many experiments have fitted this model, for example the experiments of Diamond & Solomon (1959) shown in fig. 6.4a. In such a situation the flux from the outside to the vacuole is given by $\phi_p \phi_t/(\phi_p + \phi_t)$, and ϕ_t may be got if ϕ_p is known. Often the flux to the vacuole will be rate-limited by either ϕ_p or ϕ_t; e.g. in *Chara* and *Nitella* $\phi_t \gg \phi_p$ for K^+, and $\phi_{ov} = \phi_p$. The results in fig. 6.4b do not fit the model and need further consideration.

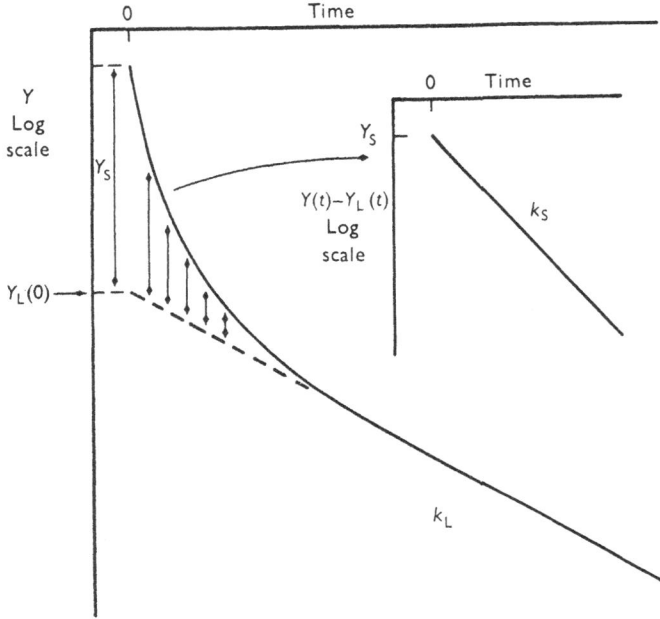

Fig. 6.3 The loss of radioactivity from a cell in an elution experiment, assuming the model in fig. 6.2. The elution results in a loss in two parts; S, short times; L, long times. The inset is the result of plotting $Y(t) - Y_L(t)$ on a log scale against time, where $Y_L(t)$ are the values of radioactivity along the dashed line, and $Y(t)$ is the total cell radioactivity at the corresponding times. Estimation of the quantities k_L, k_S, Y_L and Y_S enable calculation of the fluxes and quantities of ions. The equations are:

$$\phi_p = 1/S_0[k_S Y_S/(1 - \exp k_S t^*) + k_L Y_L/(1 - \exp k_L t^*)]$$
$$Q_c = S_0 \phi_p^2/[k_S^2 Y_S/(1 - \exp k_S t^*) + k_L^2 Y_L/(1 - \exp k_L t^*)]$$
$$\phi_t = Q_c(k_S + k_L - Q_c k_S k_L/\phi_p) - \phi_p$$
$$Q_v = \phi_p \phi_t/k_S k_L Q_c$$

where S_0 is the specific activity of the solution in which the cell was soaked for time t^*, and the other symbols have already been defined in the text.

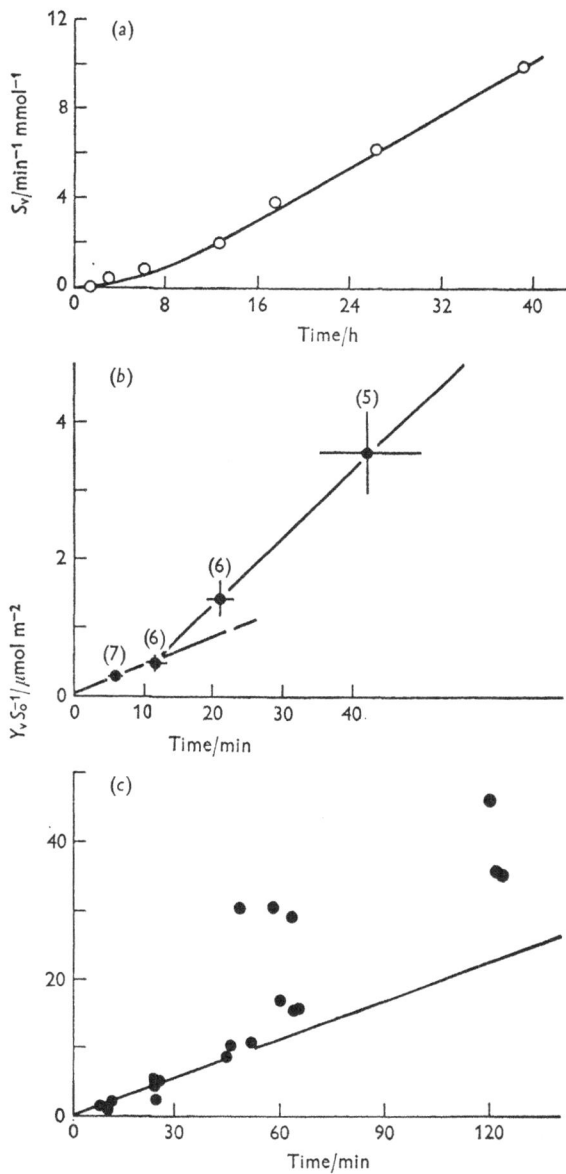

Fig. 6.4   (a) The specific activity of K$^+$ in the vacuole of cells of *N. axillaris*, as a function of time in radioactive medium (from Diamond & Solomon, 1959). (b) and (c) The entry of labelled chloride into the vacuole of cells of *N. translucens* after various times. In (c) are points for individual cells, in (b) the means and SEM over ranges of uptake times (from MacRobbie, 1969).

73

The next level of complication in modelling is to subdivide the cytoplasmic compartment into two (Larkum, 1968) or to suggest a separate pathway from outside to vacuole (Mac-Robbie, 1969). Figs. 6.5a and 6.5b show models of this kind. They will be discussed in a later section; here we may note that both can provide the feature of a very small lag time in the rise of Υ_v, needed to match the experimental results of MacRobbie (1969) shown in fig. 6.4b. These models are of some mathematical complexity, although the equations for them may be readily set up on an analogue computer. Simpler

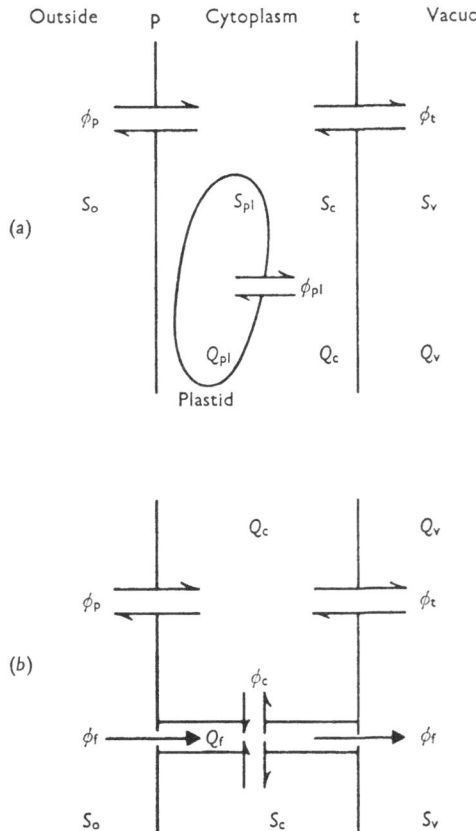

Fig. 6.5 (a) A model in which plastids (pl) form a separate compartment within the cytoplasm; influx is equal to efflux at each membrane. (b) A model with a direct flux (ϕ_f) from outside to vacuole, representing a pinocytotic vesicle mechanism. This flux goes via a phase having quantity Q_f; this phase exchanges with the cytoplasm via fluxes ϕ_c.

procedures may be used, and in particular $S_0\phi_p$ can still be identified as the initial rate of increase of cellular radioactivity, though ϕ_t becomes dependent on the details of the model. The specific activity S_c which determines the rise of vacuolar radioactivity $(\mathrm{d}Y_v/\mathrm{d}t = (S_c-S_v)\phi_t)$ becomes inaccessible to measurement in the model of fig. 6.5b and difficult to measure in that of fig. 6.5a.

Fluxes across the plasmalemma and tonoplast

Leaving aside the distinction between active and passive, we can at this stage get a picture of the rates of movement of ions in several classes of coenocytes from table 6.1. As with the electrical measurements, the data refer to cells in defined conditions, usually separated from other cells, and after some days in an artificial medium, in the light. It is seen that, generally, the fluxes of K$^+$ and Na$^+$ at the plasmalemma simply increase with their concentrations in the external medium; the fluxes are much larger in marine cells than those from fresh water. This is expected, qualitatively, if the plasmalemma fluxes of these ions are largely passive and the permeability is roughly constant. Calculation of P_K, P_{Na} is possible, for comparative purposes, if we define P_j by means of the Goldman equation (see appendix A) for the influx:

$$\overrightarrow{\phi_j} = -\frac{z_j P_j F \psi_{co} c_j^o}{RT[1-\exp(z_j F \psi_{co}/RT)]} \tag{6.1}$$

The term in square brackets is approximately unity provided ψ_{co} is more negative than $-75\,\mathrm{mV}$, which is usually true. Table 6.2 shows values of permeabilities assembled from measurements of plasmalemma influxes which were apparently not due to active transport. P_K and P_{Na} exhibit a smaller range than do the fluxes.

The tonoplast fluxes are more uncertain, but if the three-compartment model (fig. 6.2) is used, there result the large fluxes across the tonoplast given in table 6.1. That is, whether an elution analysis is made, or direct measurements made of vacuole activity, the calculated fluxes are usually of the order of 1 μmol m^{-2} s^{-1}. Some exceptions to this occur in winter cells (MacRobbie, 1966) where transfer of potassium between cytoplasm and vacuole was at a rate of 10–50 nmol m^{-2} s^{-1}, and in

TABLE 6.1 Steady state fluxes (nmol m^{-2} s^{-1}) in defined media, in the light, 20–25 °C, at the plasmalemma (p) and tonoplast (t)

Cell	$[K^+]_0$/mM	$[Na^+]_0$/mM	$[Cl^-]_0$/mM	ϕ_K	ϕ_{Na}	ϕ_{Cl}		Reference
N. translucens	0.1	1.0	1.3	8	5.5	9	p	MacRobbie (1962)
				Very high		–	t	
				12	–	10–30	p	MacRobbie (1964)
				300–1150		500–1800	t	
C. corallina	0.2	2.0	2.3	5–35	4	5–40	p	Findlay et al. (1969)
				400–1000	500–2000	–	t	
Tolypella intricata	0.4	1.0	1.6	10–20	–	20–50	p?	Smith (1968a)
Hydrodictyon africanum	0.1	1.0	1.3	14 (40)	7 (80)	14 ?	p?	Raven (1967)
Nitellopsis obtusa	0.65	30	35	2.5	4	5	p	MacRobbie & Dainty (1958)
							t	
Valonia ventricosa	12	508	596	880	35	150	'Influx'	Gutknecht (1966)
Chaetomorpha darwinii	13	500	523	650–2100	400–1100	–	p	Dodd et al. (1966)
				1000–10 000	30–40	–	t	
Chaetomorpha darwinii	10	470	550	2000	1000	2000	p	Findlay et al. (1971)
				2000	100	2000	t	
Griffithsia monilis	10	490	573	500–3800	150	70–350	p	Findlay, Hope & Williams (1970)
				>300–2000	>50	>300	t	
Valoniopsis sp.	10	470	550	500–1000	300–6000	1300	p	Findlay et al., unpublished data
				20–100	150–500	800–2000	t	
Acetabularia mediterranea	10	470	550	110–400	110–490	2000–7900	p	Saddler (1970)

TABLE 6.2 *Permeabilities to K⁺ and Na⁺ calculated from passive [a] influx at the plasmalemma*

	ϕ_{Koc}/nmol m^{-2} s^{-1}	ϕ_{Naoc}/nmol m^{-2} s^{-1}	P_K/nm s^{-1}	P_{Na}/nm s^{-1}	Reference
N. translucens	3	5.5	4.8	0.86	MacRobbie (1962)
C. corallina	4	3	6.7	0.51	Walker & Hope (1969)
	5	3	4.3	0.25	Findlay et al. (1969)
H. africanum	3	3	6.5	0.65	Raven (1967)
Nitellopsis obtusa	40	80	11	0.49	MacRobbie & Dainty (1958)
Valonia ventricosa	880	35	26	0.025	Gutknecht (1966)
	150	40	4.9	0.028	Aikman & Dainty (1966)
Griffithsia monilis	2000	150	5.8	0.089	Findlay, Hope & Williams (1970)
Valoniopsis sp.	800	>400 [b]	(80)	(>0.8)	Findlay et al., unpublished

[a] Remaining influx when Cl-linked or Na, K-linked 'active' portion is subtracted.
[b] Extremely variable amongst cells and collections.

Valoniopsis, where normally $\phi_{cv} \sim$ 40 nmol m^{-2} s^{-1}. *Valoniopsis* is not a strong accumulator of K$^+$ relative to Na$^+$ (table 5.4) and no doubt the low tonoplast flux is significant in this respect.

The kinetics of chloride movement in Nitella and Chara

The early transfer to the vacuole

As mentioned above, MacRobbie has directed attention away from the model assuming a homogeneous cytoplasm, and towards a possible vesicle mechanism in ion transport. Her evidence comes from studies of the kinetics of the distribution of chloride ions in *N. translucens*. With the advent of ^{36}Cl of higher specific activity, it became possible to examine the amount of radioactive chloride in vacuolar sap samples of *Nitella* and *Chara* cells after quite short times of immersion in radioactive solution. This was done by MacRobbie (1969), who found no time-lag in the rise of vacuolar radioactivity. Thus at 5 min, 0.15–0.35 of the activity was in the vacuolar sap and 0.85–0.65 in the cytoplasm of *Nitella*; for *Chara* we found a figure of 0.5.

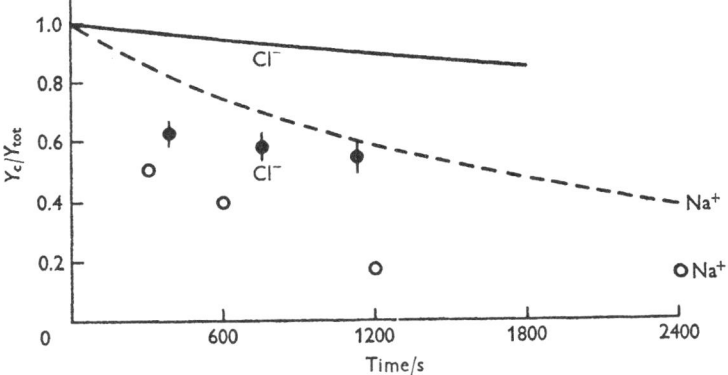

Fig. 6.6 Comparison between expectations from the simple series model, and experimental results. Filled symbols: fraction of ^{36}Cl in the cytoplasm of cells of *N. translucens* (from MacRobbie, 1969 fig. 4). Continuous line: fraction of ^{36}Cl in cytoplasm predicted by model of fig. 6.2, with $k = 2 \times 10^{-4}$ s^{-1}. Open symbols: fraction of ^{24}Na in the cytoplasm of cells of C. *corallina*, from Findlay *et al.* (unpublished). Dashed line: fraction of ^{24}Na in cytoplasm predicted by model of fig. 6.2, with $k = 10^{-3}$ s^{-1} (the largest value likely).

The fraction in the cytoplasm expected from the simple model (fig. 6.2) is $[1 - \exp(-kt)]/kt$ where k is the cytoplasmic rate-constant for chloride, equal to $(\phi_p + \phi_t)/Q_c$. The expected fraction from the simple model would appear to be 0.93 (*N. translucens*) or 0.86 (*C. corallina*) on the basis of mean fluxes and the quantity of chloride for the whole cytoplasm (tables 6.1, 5.5).

The fraction in the cytoplasm changes with time in the way shown in fig. 6.6, where a theoretical prediction from the simple model is also shown. The discrepancy between experiment and theory here represents an unexpectedly large vacuolar radioactivity at short times. The inference is that the radioactive chloride reaching the vacuole does not mix with the whole of the cytoplasmic chloride, as was assumed in the theory. The agreement between experiment and the prediction of the simple model at later times implies that a larger pool of cytoplasmic chloride has come to have a specific activity equal to that of the small pool which is labelled at short times.

Each of the models of fig. 6.5 can be fitted to data of this kind We shall consider how far that of fig. 6.5*a* can be made to fit. The chloroplasts can be regarded as a separate cytoplasmic compartment containing for *Chara* say 1 mmol Cl$^-$ m^{-2} (table 5.5) with fluxes ϕ_{pl} between flowing cytoplasm and phase pl (for 'plastids'). If the flowing cytoplasm contains 30 μmol m^{-2} Cl$^-$ (10 mM Cl$^-$ in a layer 3 μm thick – see table 5.5) it is only necessary, to match the observed kinetics, to adopt different light and dark values for ϕ_p and ϕ_t (fig. 6.7). The fit of the model to these observations on *Chara* seems reasonable, and the *Nitella* results can be fitted in the same way. MacRobbie (1969) rejected the model with plastids and flowing cytoplasm as separate compartments, partly on the grounds that measured values of Q_c were of the order of 1 mmol m^{-2}. This argument simply ignores the lower measured values of table 5.5 which, as we saw, allow the rejected model to fit the kinetic data.

Other features of the chloride kinetics have been reported (MacRobbie, 1969): (a) the apparent flux into the vacuole ($\Upsilon_v/S_o t$) at short times is directly correlated with ϕ_p, the influx into the whole cell across the plasmalemma, when batch means are compared, and (b) at later times, when k is about constant, k is nevertheless proportional to ϕ_p, when batch means are examined. In both (a) and (b) the different means were obtained by changing temperature, chloride concentration, age of cells, by adding CCCP, and so on. The doubtful significance of correlations obtained in this way does not allow much argument to depend on them. In particular neither flux can be said to control the other on such evidence.

The situation for potassium and sodium ions has not been as fully investigated but it appears that there may be a 'rapid uptake' of sodium (fig. 6.6), while there may be no such effect for potassium (MacRobbie, 1969).

Other cells have also revealed unexpected relationships between vacuolar and plasmalemma fluxes. In *Griffithsia* (Findlay, Hope & Williams, 1970), ϕ_t was proportional to ϕ_D for potassium ions. However, there was strong evidence that ϕ_D was mostly a passive flux, and a control of vesicle formation by a passive flux seems unlikely. Most of the electrical evidence from studies of *Griffithsia* and also *Chaetomorpha* is in agreement with the plasmalemma and the tonoplast presenting series resistances to ion movements, with the cytoplasm as a compartment separating the medium from the vacuole. However, some of the flux data would lead to the tentative conclusion that

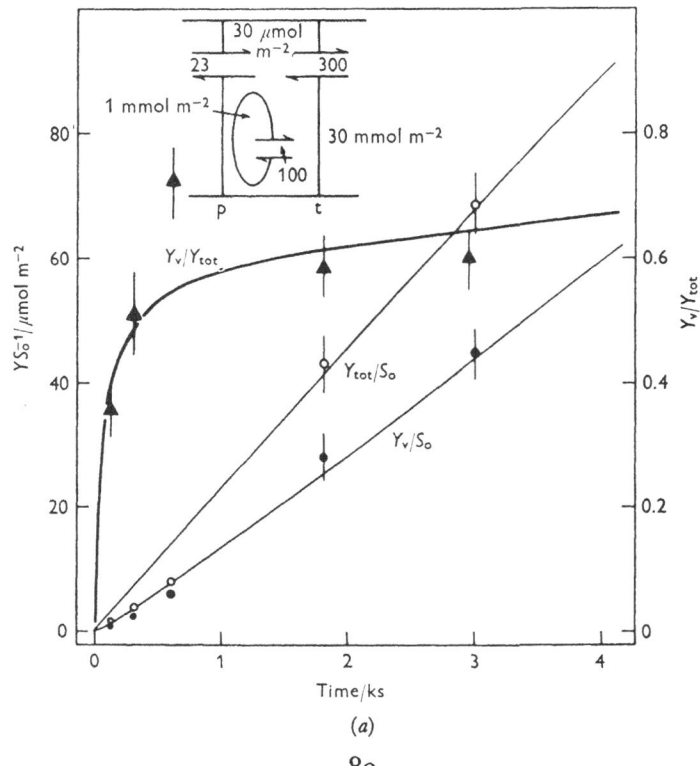

(a)

there are parallel paths from the medium to the vacuole and to the cytoplasm. Permanent long invaginations of the plasma-lemma visible in some electron micrographs of animal cells do not appear to occur in plant cells. While vesicles that form at the plasmalemma and discharge at the tonoplast may well be the answer to some of the perplexing observations we have been discussing, a quantitative treatment is needed to show how such a model might behave in detail. Its predictions could then be tested.

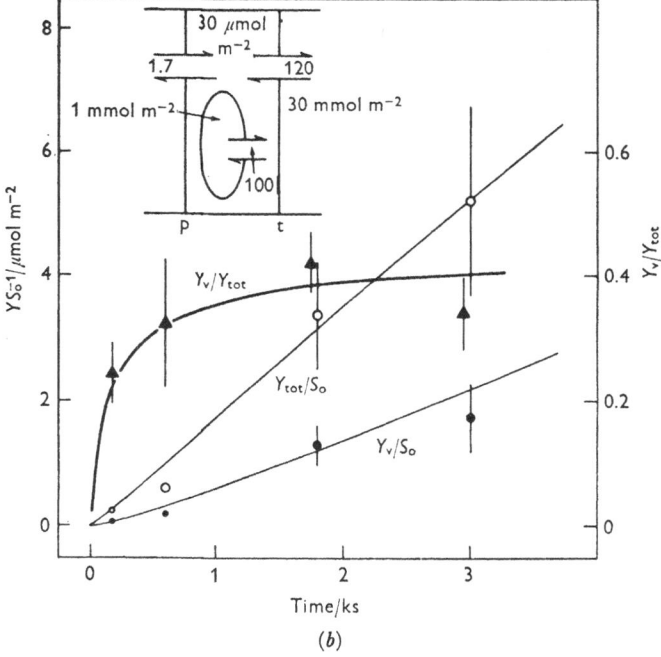

Fig. 6.7 (a) Calculated values of Y_v/S_o, Y_{tot}/S_o and Y_v/Y_{tot} obtained from an analogue computer set up to model the arrangement of compartments shown in the inset (cf. fig. 6.5a). The points plotted are results for chloride uptake by *C. corallina* in light (Findlay *et al.*, unpublished). (b) As for (a) except that the model and results are for cells in the dark.

The distribution of the fraction reaching the vacuole at short times

An extension of the work with *N. translucens* led MacRobbie (1970a, b) to state that 'in the individual cells in any given experiment the rate of transfer to the vacuole in the fast component is a quantized fraction of the total influx'. By 'transfer to the vacuole in the fast component' is meant the unexpected early appearance of radioactive chloride in the vacuole. Only in terms of a particular model (with a 'straight-through' path fig. 6.5b) can this validly be described as a 'component' of the total flux to the vacuole. The hypothesis of quantisation is that early values of the fraction of radioactive chloride in the vacuole are not, within a batch of similar cells, randomly scattered in a featureless distribution; but that they are scattered close to values of $n\alpha$ where n is an integer 1, 2, 3 . . ., and α is a constant within each batch, varying strikingly (0.03–0.3) between batches. This hypothesis is not supported by the data nor by valid statistical tests (see Findlay, Hope & Walker, 1971; Walker, 1973) and is not furthered by later arguments (MacRobbie, 1973).

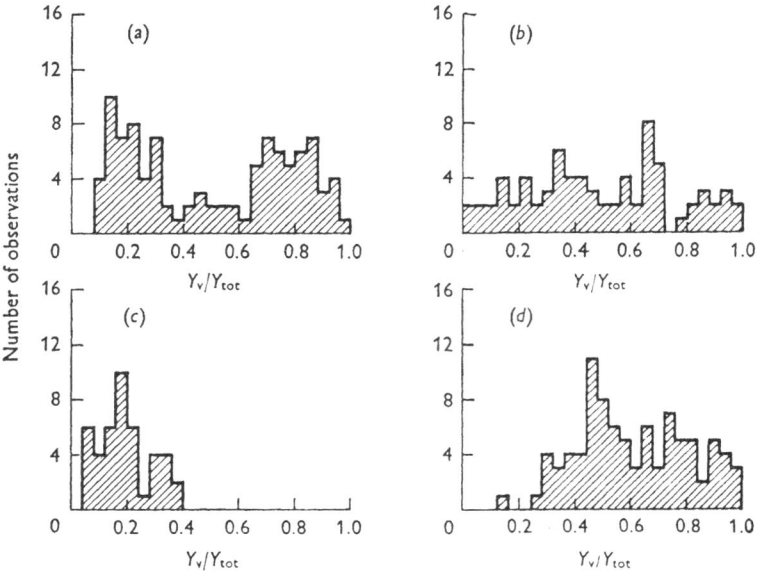

Fig. 6.8 Histograms showing the distribution of values of the fraction of chloride in the vacuole after short times: (a) and (b), *C. corallina*, uptake time 600 s, temperature 22–25 °C; (c) *N. translucens*, experiment NA3 from MacRobbie (1970a); (d) *C. corallina*, uptake time 600 s, 9 °C. The interval in each histogram is 0.04. (a), (b) and (c) are from Findlay, Hope & Walker (1971), (d) from unpublished results of Findlay *et al*.

In experiments with *Chara*, the distributions for $\Upsilon_v/\Upsilon_{tot}$ found when about 100 cells have been examined under conditions that should have revealed any grouping, have not shown any quantisation (fig. 6.8). No significant grouping was found by Findlay (unpublished data, 1972) in a 100-cell experiment with *Tolypella*, the same species which in one experiment (MacRobbie, 1970a) had shown some grouping effect. Thus, at present, the phenomenon of quantisation seems to have no firm experimental basis.

The very wide spread in values of $\Upsilon_v/\Upsilon_{tot}$ (and of ϕ_p) in experiments such as the ones described is baffling, and an impediment in experiments which seek to examine the effects of treatment on fluxes. Many questions indeed remain unanswered:

(i) How well does efflux or 'elution' data fit the new models?

(ii) Action potentials may occur at the tonoplast (MacRobbie, 1973 – see chapter 8): how do these affect the distribution of radioactive label found by techniques that involve cutting excitable cells such as the charophytes?

(iii) What is the real status of evidence used to support ideas of vesicular transport within the cytoplasm? The unlikeliness of quantisation, the promising match between the four-compartment model of fig. 6.5a with some data, (fig. 6.7) and the inconclusive nature of certain correlations, all discussed above, lead us to a position of scepticism.

D

CHAPTER 7

Electrical properties of membranes

Potential differences in giant algal cells

We have seen earlier in this account how giant cells attracted electrophysiologists in the 1920s and 1930s. This burst of electrical activity reflected the discovery of certain of the giant genera and the ideas of Loeb and Bernstein that biological phenomena had a physico-chemical basis. In the 1960s there was renewed enthusiasm, inspired by the example of the animal physiologists, when algae, mosses and roots, coleoptiles, and leaves of higher plants were stuck with electrodes.

All cells so far measured are electrically negative in the cytoplasm relative to the external medium. Location within the cytoplasm does not seem to matter, to a first approximation. In terms of the models of chapter 6, workers with microelectrodes have always used that shown in fig. 6.2, in which the cytoplasm is all one phase. The vacuole may be found to be positive or negative with respect to the medium. Table 5.4 summarises the main findings with coenocytes.

The potential difference between the cytoplasm and the medium may not reflect the p.d. across the plasmalemma itself because of the ionic nature of most cell walls, but some of the properties of the plasmalemma can be inferred from such measurements if the cell wall is in thermodynamic equilibrium with the medium or if the exact ionic composition of the wall is known under various conditions. Unfortunately, the condition of equilibrium is not obviously satisfied in many studies, because of the slow exchange of the wall counterions, especially the divalent ones (see chapter 5).

The effect of the ion-exchange properties of the wall is to introduce steps in potential and in average ion activities between medium and plasmalemma. It is only possible to calculate these steps approximately because of unknown factors in the structure of the wall. The wall ion-exchange system and plasmalemma could be physically in close apposition, or there may be a gap between them filled with uncharged cellulose material. This is illustrated in fig. 7.1 which also shows the approximate steps in p.d. and ion activity expected with the fresh water type of coenocyte such as *Nitella*. It can be seen, therefore, that what is reported as a 'membrane' potential may, in the situation of fig. 7.1*b*, actually be the sum of two p.d.s, that across the plasmalemma being only one of the two components. The

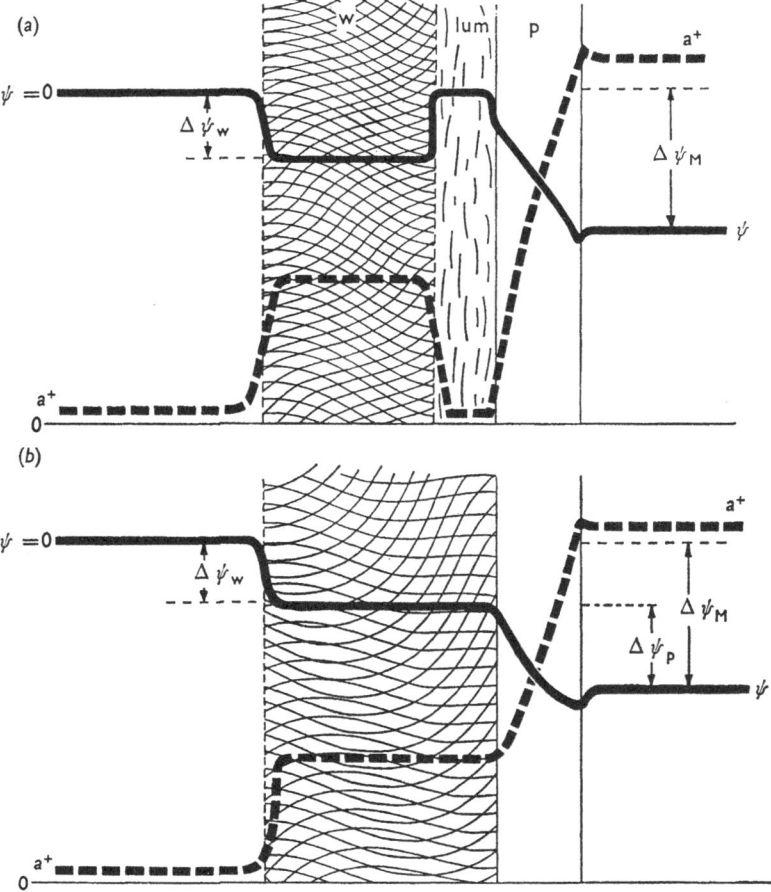

Fig. 7.1 Schematic diagrams showing possible profiles of potential (ψ) and cation concentration (a^+) in cell wall (w) and plasmalemma (p): (*a*) if there were an uncharged lumen (lum) between w and p, and (*b*) if the wall were homogeneously charged throughout. The wall is assumed to have a net negative charge. $\Delta \psi_M$ represents the observed membrane p.d., $\Delta \psi_w$ the wall p.d. and $\Delta \psi_p$ the plasmalemma p.d.

effect of this on conclusions made from studies of the p.d. has been discussed by Hope & Walker (1961), Spanswick, Stolarek & Williams (1967) and Vorobiev, Radenovich, Khitrov & Yaglova (1967), and will be summarised below.

Attempts to measure a 'wall p.d.' were made by Nagai & Kishimoto (1964) who inserted microelectrodes into the cell wall. In so far as the 3 M KCl inside the electrode is concentrated enough to reduce the Donnan p.d. at the tip to a low value, their method gives approximately the p.d. between wall and medium. Their values are in general agreement with those in table 5.1. The potentials they observed in the outside medium up to 20 μm from the wall are artifacts due to the diffusion of KCl from the electrode tip; ionic double layers are three orders of magnitude narrower than this. The measurements of wall p.d. by Spanswick, Stolarek & Williams (1967) show elegantly the dependence of the p.d. on $[Ca^{2+}]_o$ and its independence of $[K^+]_o$ and $[Na^+]_o$ as would be expected for a Donnan p.d.

Ion permeability ratios inferred from p.d.'s

Since the wall ion-exchange material was not expected to distinguish between K^+ and Na^+, the changes in p.d. produced by changes in $[K^+]_o$ (keeping $[K^+]_o+[Na^+]_o$ constant) were taken by Hope & Walker (1961) as reflecting changes in plasmalemma p.d.

In this study of *Chara*, and in *Nitella* (Spanswick *et al.*, 1967) a peculiar effect was observed. In media containing Ca^{2+}, the plasmalemma p.d. is not very dependent on $[K^+]_o$ except at high concentrations: see fig. 7.2a. If the cells are soaked for some hours in 5 mM NaCl, and then measured in solutions containing no Ca^{2+}, the plasmalemma p.d. is dependent on $[K^+]_o$ over a wide range of concentrations (fig. 7.2b). This dependence is reasonably well expressed by the equation:

$$\psi_{10} = (RT/F) \ln \{([K^+]_o + \alpha[Na^+]_o)/([K^+]_1 + \alpha[Na^+]_1)\} \quad (7.1)$$

(see appendix A).

The inference was drawn that the membrane p.d. is under these circumstances determined by the diffusion of K^+ and Na^+ down their electrochemical gradients. Briggs (1962) showed that an electrogenic pump delivering a constant current would have passed undetected in these experiments. The values of α, the ratio of sodium permeability to potassium permeability, were about 0.1 for *Chara* and 0.25 for *Nitella*.

In the untreated cells only small changes of plasmalemma p.d. occur as $[K^+]_o$ is increased, until at some concentration between 1 and 10 mM, an action potential occurs, the membrane remains somewhat depolarised, and now behaves like the membrane of a pre-treated cell. This description implies that

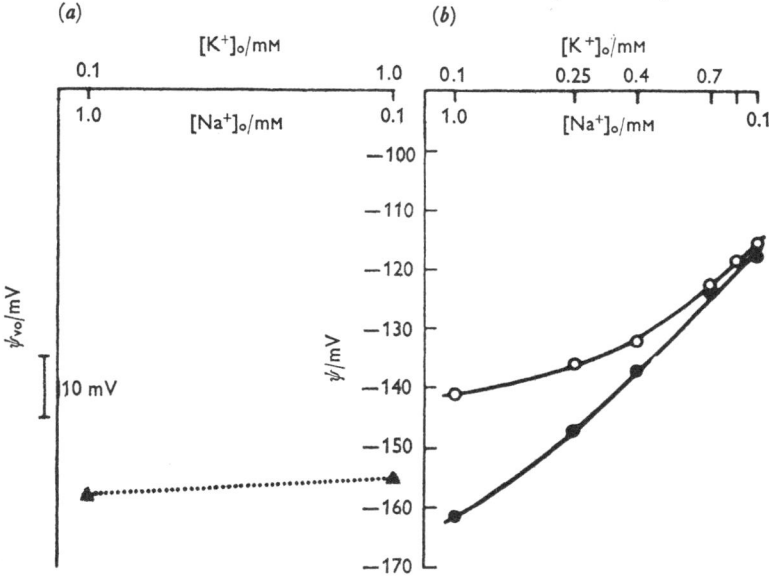

(a)

(b)

Fig. 7.2 (a) Electric potential difference measured between the vacuole and external medium, showing the small effect of changing $[K^+]_o$, when Ca^{2+} was in the medium: *N. translucens*, replotted from Spanswick, Stolarek & Williams (1967); only changes in p.d. were recorded, hence the absence of axis values. (b) Changes in p.d. when $[K^+]_o$ and $[Na^+]_o$ are changed, keeping $[K^+]_o + [Na^+]_o = 1.1$ mM, after pretreating cells in a calcium-free medium. (●), *C. corallina* from Hope & Walker (1961); (○), *N. translucens* from Spanswick, Stolarek & Williams (1967). The lines are solutions of text equation (7.1) with α = 0.07 or 0.256 and $[K^+]_i + α[Na^+]_i = 105$ or 100 mM, for *C. corallina* and *N. translucens* respectively.

in solutions of $[K^+]_o$ near 1 mM, the membrane p.d. is more negative than the potassium equilibrium potential – see fig. 7.3 (Oda, 1962). Transition from the depolarised to the hyperpolarised state can be produced by reduction of $[K^+]_o$ or by applied electric current (Oda, 1962; Kishimoto, 1966). The source of this hyperpolarisation is not clear, the contending explanations suggesting the presence of an electrogenic active transport of Cl⁻ inwards or of H⁺ outwards (Kitasato, 1968; Spanswick, 1973). Before discussing this question we will consider the results from the marine algae: *Griffithsia, Chaetomorpha* and *Acetabularia* have been the most thoroughly studied. While it has been concluded that α is about 0.1–0.3 in the

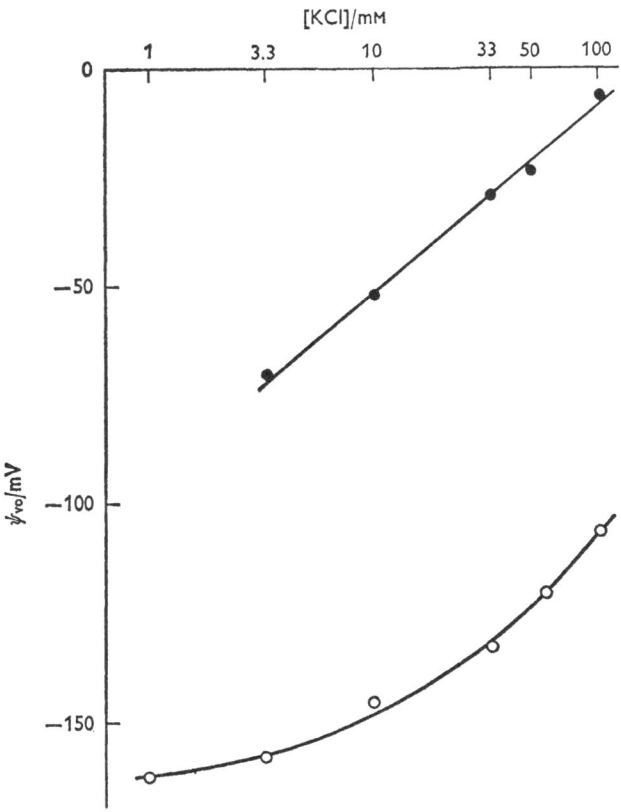

Fig. 7.3 Resting p.d. before (O) and after (●) the occurrence of an action potential, as a function of [KCl] in the medium: *C. braunii* (replotted from Oda, 1962).

charophytes, it is much smaller in the marine genera. Thus in *Griffithsia* it is 0.002 to 0.004, in *Chaetomorpha* 0.001 to 0.01 and in *Acetabularia* 0.02 to 0.06. Generally, the internal concentration of potassium calculated from the results is a reasonable approximation to the measured concentration of potassium in the cytoplasm. Thus in many cells the p.d. across the plasma-lemma is effectively set by the ratio $([K^+]_o+\alpha[Na^+]_o)/([K^+]_c+\alpha[Na^+]_c)$. In others, especially *Chaetomorpha* and *Valoniopsis*, and probably *Valonia*, the p.d. is sensitive to changes in $[Cl^-]_o$. The simplest way to accommodate a passive permeability to Cl^- is to use an equation formally similar to (7.1) but based on the assumption that the electric field in the membrane is constant (Goldman, 1943). This equation is

$$\psi_{co} = \frac{RT}{F} \ln \frac{[K^+]_o+\alpha[Na^+]_o+\gamma[Cl^-]_c}{[K^+]_c+\alpha[Na^+]_c+\gamma[Cl^-]_o} \qquad (7.2)$$

where γ is the permeability ratio $P_{Cl}:P_K$ (see appendix A). γ has not been evaluated with much accuracy, but was about 0.03 in some experiments with *Chaetomorpha*. Changes in p.d. for three marine genera are shown in fig. 7.4. In *Griffithsia*, equation (7.1), implying passive movements of K^+ and Na^+, seems to describe the p.d. adequately. In *Chaetomorpha* a passive permeability to Cl^- is postulated (equation 7.2). In *Acetabularia* two membrane states occur, of which the hyper-polarised (at 25 °C) is attributed by Saddler (1970) to the outward current carried across the membrane by an inward electrogenic chloride pump. In each state passive permeability to K^+ and to Na^+ affects the p.d.; in the hyperpolarised state

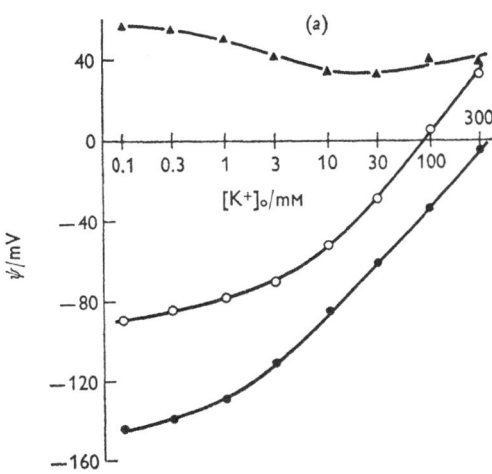

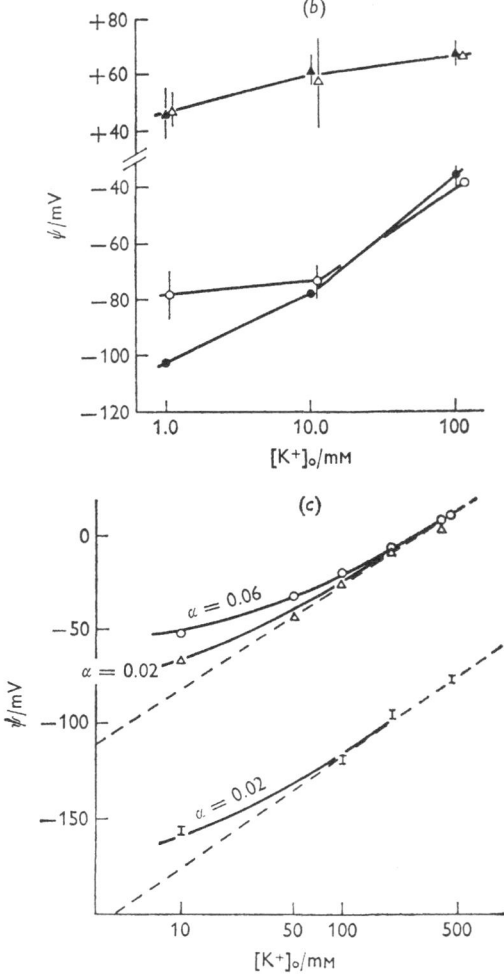

Fig. 7.4 (a) $\psi_{vo}(\bigcirc)$, $\psi_{co}(\bullet)$ and $\psi_{vc}(\blacktriangle)$ plotted against $[K^+]$ for *Griffithsia pulvinata*. The curves through ψ_{vo} and ψ_{vc} were drawn by eye, while that through ψ_{co} is based on text equation (7.1) with $\alpha = 0.002$, $[K^+]_o + \alpha[Na^+]_o = 350$ mM. The SEM for all points was less than 3 mV (from Findlay, Hope & Williams, 1969). (b) The effect of changing $[K^+]_o$ on $\psi_{co}(\bigcirc, \bullet)$ and ψ_{vc} ($\triangle$, $\blacktriangle$) in *Chaetomorpha darwinii*, for two values of $[Cl^-]_o$, 573 mM ($\bullet$, $\blacktriangle$) and 191 mM ($\bigcirc$, $\triangle$) (from Findlay *et al.*, 1971). (c) The effect of $[K^+]_o$ on the membrane potential at 5 °C ($\bigcirc$, $\triangle$) and 23 °C (lower curve) in *Acetabularia mediterranea*. The dashed lines have slope 56 (5 °C) and 59 (23 °C) mV per tenfold change in $[K^+]_o$. The curves were obtained by using text equation (7.1) with α as shown, and $[K^+]_i + \alpha[Na^+]_i = 290$, 350 mM ($\bigcirc$, $\triangle$). The curve for 23 °C is similar but has a constant value of -85 mV added (from Saddler, 1970).

ψ_{co} is more negative than the Nernst equilibrium potential for any ion. This occurs in *Chara* as well. The Nernst potential for ions j (see appendix A) is given by:

$$\psi_j = (RT/z_j F) \ln (c_j^o/c_j^c) \qquad (7.3)$$

The permeability ratios referred to are not to be regarded as fundamental constants for any particular genus. They may sometimes be constant over a range of values of ψ_{io} or of concentration but more often they are not constant. Frequently their value depends on $[Ca^{2+}]_o$, e.g. in *Chaetomorpha*. A comprehensive theory for membrane p.d.s has not yet been developed that will cope with this apparent dependence of the permeability on concentration or ψ.

Apart from K^+, Na^+ and Cl^-, the only other ion species suggested to permeate the plasmalemma readily is H^+ (or H_3O^+) (Kitasato, 1968). This will be considered, together with the question of electrogenic H^+ pumps, after the discussion of membrane resistance.

Potential difference across the tonoplast

The potential difference observed between the vacuole and the cytoplasm in giant coencytes may have a value between about 10 mV and 80 mV, the cytoplasm being always negative with respect to the vacuole. The tonoplast p.d. is almost equal and opposite to the plasmalemma p.d. in *Chaetomorpha* and *Valonia*, so that the p.d. between vacuole and medium is ± 10 mV. In charophyte cells on the other hand the vacuole and cytoplasm differ in potential by only 20–25 mV and ψ_{vo} is sometimes taken as an easy measure of ψ_{co}. It is possible to change the ionic concentrations on the inside of the tonoplast by a perfusion technique (Blinks, 1935; Tazawa, 1964; Gutknecht, 1967a) but this has not revealed a great deal about relative permeabilities of the tonoplast to ions, the p.d. being relatively insensitive to ion changes. The charophyte tonoplast has a p.d. not too far from the Nernst equilibrium potential for K^+. In *Chaetomorpha* the tonoplast p.d. has two characteristic values, about 40 or 80 mV, the latter apparently being close to the equilibrium potential for chloride. Spontaneous transitions from one value to the other are observed.

91

Electric conductances

A survey of electric conductance measurements on the two membranes of giant algal cells is given in table 7.1. In freshwater cells, the conductance of the plasmalemma is generally the lower of the two membranes; it largely determines the conductance of the whole protoplast. Since the conductance of the plasmalemma generally depends on the ionic composition of the medium, more attention has been given to it than to the more difficult tonoplast conductance.

Although in the thirties and forties there were attempts to understand the plasmalemma conductance in terms of 'polarisability' or of the movement of electrons, it has become accepted that the appropriate model is given by

$$g_M = \sum_j g_j \qquad (7.4)$$

where g_M is the measured membrane conductance (reserving the symbol g for differential conductance $\partial \mathcal{J}/\partial \psi$) and g_j are the partial ionic conductances. This is generally taken to imply that the mechanism of conduction of electricity which gives rise to the conductivity is the passive movement of ions in response to electrochemical gradients. The investigator's task was accepted as being the identification of the ions j which contribute significantly to the sum, and the elucidation of the way in which g_j depends on the concentrations and the potential difference.

We will consider first how far this approach has been successful, before discussing how it is being modified. In what follows, the summation is taken over K^+, Na^+ and Cl^- only. Ionic conductances may be most simply related to ion concentrations and potential differences by using a set of equations based on Goldman's assumption of a constant electric field in the membrane. The chord conductance, that measured in practice from the change in current producing a stepped change in p.d. is given by:

$$G_M(\Delta \psi) = \Delta \mathcal{J}_M/\Delta \psi_M = \Sigma \Delta \mathcal{J}_j/\Delta \psi_M = \Sigma G_j(\Delta \psi_M) \qquad (7.5)$$

and:

$$G_j = \Delta \mathcal{J}_j/\Delta \psi_M = \frac{P_j c_j^0 F^2 \psi_M [1 - \exp(F\Delta \psi_M/RT)]}{RT \Delta \psi_M [1 - \exp(F\psi_M/RT)]} \qquad (7.6)$$

(see appendix A).

TABLE 7.1 *Membrane Conductance* ($S\ m^{-2}$) *in giant algal cells*

Cell	Medium	g_p	g_t	g_{tot}	Comment	Reference
Nitella sp.	White's soln [a]			0.28		Findlay (1959)
Nitella sp.	KCl 1 [b]	1.5	>3			Walker (1960)
	NaCl 1	0.7				
	CaCl 0.5	0.5				
C. corallina	APW [c] (K+ 0.1)			0.6	External contacts	Walker (1960)
	K+ 0.1–1.9			0.7	Microelectrode method	Hope & Walker (1961)
				0.6–2.5	Conductance largely dependent on $[K^+]_0$ if cell p.d. responded also to $[K^+]_0$	
C. corallina	APW	0.8	10	–		Findlay & Hope (1964)
C. braunii	0.05 K+, etc.			0.33		Oda (1961)
N. translucens	APW			0.4		Williams, Johnston & Dainty (1964)
N. translucens	APW			0.35		Hogg, Williams & Johnston (1969)
N. translucens	APW	0.11	1.2	–		Spanswick & Costerton (1967)
Nitellopsis obtusa	AHW	0.12	1.0	–		Findlay (1970)
Nitellopsis obtusa	APW	–	–	0.29		Paszewski, Stolarek & Gebal (1968)
Griffithsia spp. [d]	ASW	50	2	–		Findlay, Hope & Williams (1969)
Chaetomorpha darwinii	ASW	20	1.9	(1.7)		Findlay *et al.* (1971)

[a] $[K^+] = 0.17.$ [b] Conc., mM. [c] See table 5.1. [d] *G. monilis, G. pulvinata, G. teges.*

We can calculate permeabilities P_j from this set of equations if we have values for G_M and for the permeability ratios α and γ for the important permeant ions K$^+$, Na$^+$, Cl$^-$. These permeabilities would have the significance that they should represent the product of partition and mobility coefficients, if the equation (7.6) were soundly based. They are, however, best regarded as empirical parameters in a conveniently simple equation. As we shall see, there are more urgent problems than the basis of equation (7.6).

For *Chara*, the calculation discussed yields the values

$$P_K = 1.4 \times 10^{-7} \text{ m s}^{-1}; \qquad P_{Na} = 1.4 \times 10^{-8} \text{ m s}^{-1}:$$

and for *Griffithsia*,

$$P_K = 3.0 \times 10^{-7} \text{ m s}^{-1}; \qquad P_{Na} = 2.2 \times 10^{-9} \text{ m s}^{-1}.$$

The permeabilities referred to above are usually calculated from 'resting' conductances, obtained from a small change in potential (5–10 mV) upon the application of a small current (0.5–1 mA m^{-2}). Further information has been obtained from a study of a wider range of currents, that is, by obtaining the dependence of ψ_M on J. Such a curve is usually asymmetrical – the membrane displays rectification, of the kind shown in fig. 7.5a for *Tolypella* (Fujita, 1962). Equation (7.6) implies rectification, in that each G_j is a function of ψ_M and of $\Delta\psi_M$. Hope & Walker (1961) found reasonable agreement between these predictions and the variation of conductance with potential in *Chara* (fig. 7.5b). However, Coster (1965) discovered an unexpected phenomenon which is difficult to explain with the simple model of the plasmalemma so far advanced: a model which assumes permeability coefficients which depend on partition and mobility coefficients but not on membrane potential. The phenomenon is 'punch-through', a sharp rise in membrane conductance at a p.d. of about −300 mV in *Chara*, that is, at a hyperpolarisation of about −150 mV. Coster (1969) showed that the punch-through potential depended on the pH of the medium. The slope conductance g_M goes to infinity at a finite membrane potential which depends sharply on pH$_0$ (fig. 7.5c). Coster & Hope (1968) showed a marked rise in chloride conductance under similar conditions which presumably underlies the effect. To

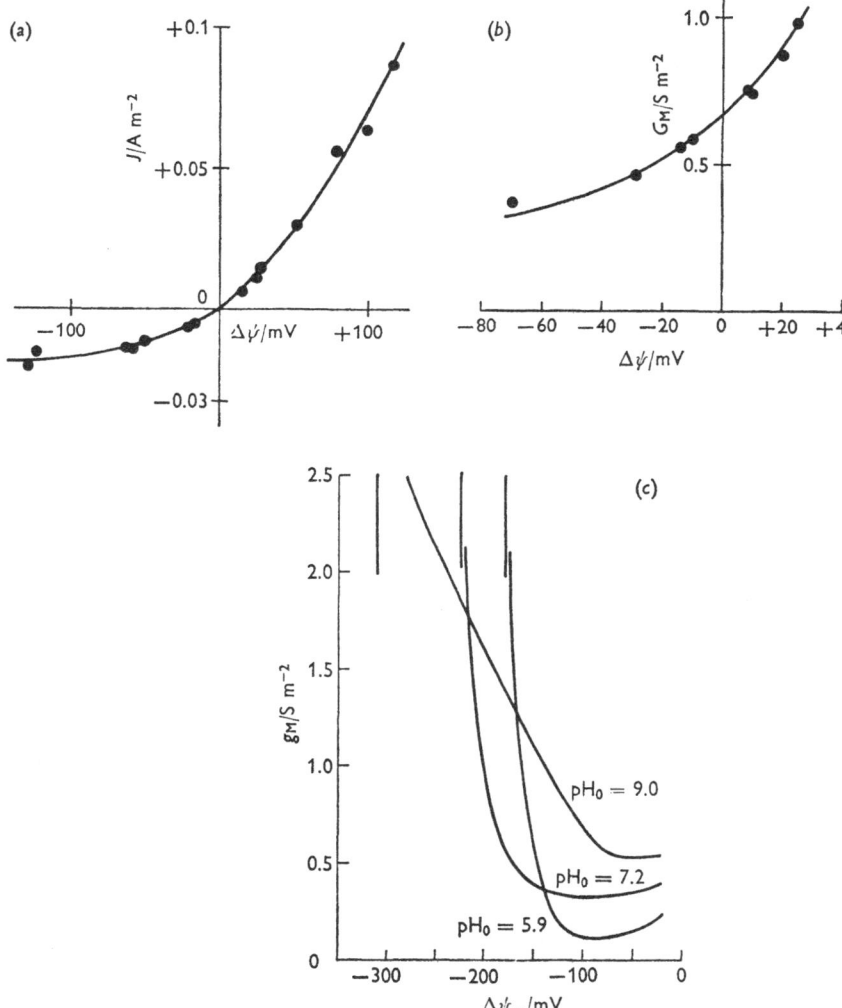

Fig. 7.5 (a) Current: voltage curve for a cell of *Tolypella gracilis* (replotted, from Fujita, 1962). A positive $\Delta\psi$ means depolarization and positive current density means an outward current of positive ions. (b) Chord conductance ($G_M = \Delta J/\Delta\psi$) measured between vacuole and medium for a cell of *C. corallina*. The curve through the points is based on text equations (7.5) and (7.6) (replotted, from Hope & Walker, 1961). (c) Slope conductance ($g_m = \partial J/\partial\psi$) as a function of change in ψ_{vo}, for three different values of pH in the medium, for cells of *C. corallina*. The conductance became infinitely large at the asymptotes shown (from Coster, 1969).

95

explain the strong dependence of permeability on membrane potential Coster has adopted the model of two fixed-charge layers of opposite sign in close contact. The contact induces a space charge region or depletion layer that changes in width with the applied potential; when the latter is made large enough, the depletion layer approaches the total thickness of the membrane, upon which there is an avalanche of current – as in the Zener diode. The effect of changed pH is presumably on the charge density in one or both of the fixed-charge layers (Coster, 1969). This model has also been used to describe the membrane resting potential as a function of ion concentrations, and its impedance at low frequencies (Coster, 1973a, b, and see below).

Fluxes and conductances

Readers will have already noted the disturbing fact that the permeabilities calculated above from electric conductances, and those from ionic fluxes (table 6.2), are different by more than an order of magnitude. Admittedly the discrepancy could be due to the use of an unsatisfactory theory to calculate the permeability parameters, but to many minds the missing consideration is whether some other ions besides K^+ and Na^+ and Cl^- contribute to yield a larger membrane conductance. The first possibility was often canvassed when the discrepancy was first noticed. The equation (7.6), and that from which the fluxes are expressed as permeabilities (6.1), both rely not only on the constant field assumption, but on the assumption of passive, independent ion diffusion. Hodgkin & Keynes (1955) have shown that for K^+ in the giant axon of *Sepia*, this assumption does not hold. Here, the potassium partial conductance is about 2.5 times greater than the value calculated from its flux, assuming passive, independent diffusion. This factor, presumably due to self-interaction of K^+ ions as they cross the membrane, was invoked in the *Nitella* and *Chara* puzzle. The factor had, however, to be ten times, rather a lot of self-interaction; this was in fact no more than an *ad hoc* assumption.

The discrepancy can be expressed in a more direct way by attempting to calculate the expected conductance directly from the measured flux. This is possible from equations related to (6.1) and (7.4) for passive, independent diffusion. In the special case of an ion in equilibrium, the relation is quite simple (Hodgkin, 1951):

$$g_j = z_j^2 F^2 \phi_{joc}/RT \qquad (7.7)$$

where ϕ_{joc} ($= \phi_{jco}$) is the unidirectional flux (see appendix A). This gives, for a flux of 10 nmol m^{-2} s^{-1} ($\equiv$ 1 pmol cm^{-2} s^{-1}) a partial conductance of 0.04 S m^{-2} ($\equiv 4$ μmho cm^{-2}). The calculation may be done for the passive K$^+$ fluxes in *Chara* and *Griffithsia* with the following results:

	ϕ_K/nmol m^{-2} s^{-1}	g_K/S m^{-2}	g_M/S m^{-2}
C. corallina	5	0.02	0.8
G. monilis	2500	10	70

In each case K$^+$ is apparently by far the dominant passive ion – for example the calculated g_{Na} for *C. corallina* is 0.006 S m^{-2}, which is no help. Similarly, conductances of calcium and of chloride do not add appreciably to the sum.

Empirical estimations of the partial ion conductances in *Chara* were attempted (Walker & Hope, 1969), by measurement of influx and efflux of K$^+$ and Na$^+$ as functions of ψ_{co}. This procedure leads to g_K, g_{Na} through the relation:

$$g_j = F \partial \phi_j / \partial \psi_{co} \qquad (7.8)$$

where $\phi_j = \phi_{joc} - \phi_{jco}$. The experiments involved making a series of measurements of fluxes into and out of cells into which electrodes had been inserted to measure and control ψ by means of voltage clamping. The protracted nature of the experiments forces one to make vacuolar recordings only, since electrodes in the cytoplasm become sealed off too soon. The possibility of making such experiments is confined to freshwater plants, so far, because the passage of a current from the vacuole of marine cells such as *Griffithsia* changes the tonoplast p.d. much more than the plasmalemma. This makes it virtually impossible to examine ϕ_{oc} as a function of ψ_{co} if ψ_{co} is to be changed by a current.

The prolonged passage of current during voltage clamping has other effects besides that of changing ψ_{co}; for example ion concentration profiles in the cell walls may well be altered. This is strongly suggested by an exponential decline of the current required to hold the membrane potential at a value different from the resting. This decline has a halftime of about 90 s.

This and other effects lead to underestimation of changes in the influx and efflux. If some means of reducing these errors can be devised, it will be worth repeating these experiments, of which fig. 7.6 is a summary for K^+ in *C. corallina*. The partial conductances g_{Cl} and g_{Na} were estimated from measurements on their efflux and influx respectively.

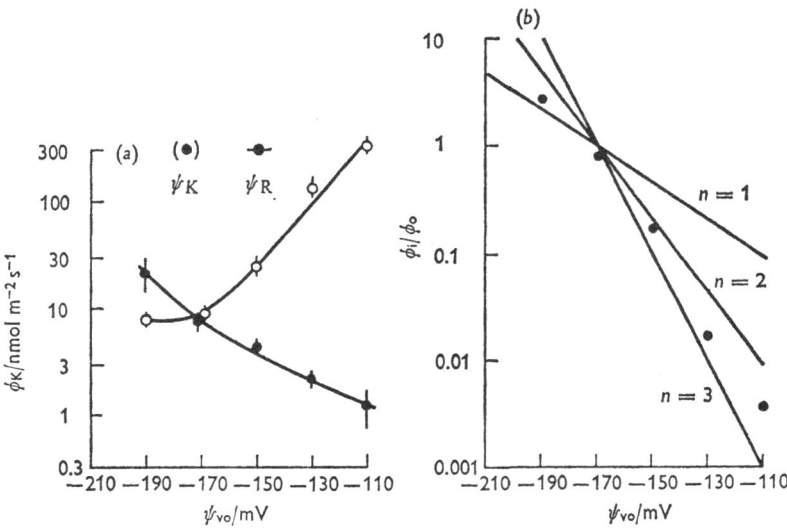

Fig. 7.6 (*a*), mean influx (●, 10 cells) and efflux (○, 7 cells) of K^+ in *C. corallina* in APW, plotted against ψ_{vo}, data from voltage clamping experiments. The fluxes and p.d.'s were grouped in intervals of p.d. such as −100 to −119 mV (plotted as −110 mV), etc. ψ_R is resting p.d., ψ_K the Nernst p.d. for K^+. (*b*) flux-ratios for K^+, using the mean fluxes on the left diagram, plotted against ψ_{vo}. The lines are theoretical expectations from an equation similar to (7.9) (from Walker & Hope, 1969).

There are several conclusions from these results that are of fundamental interest.

(i) The partial conductance for K^+ was 0.15 S m^{-2} at the resting potential. The average conductance of the membrane was 0.7 S m^{-2}. Even with estimated contributions of 0.01 and 0.02 S m^{-2} from Na$^+$ and Cl$^-$, the discrepancy is still a factor of 4, with $g_M > \sum_j g_j$.

(ii) The variation of flux-ratio with ψ is much greater than expected from independent diffusion. The results fit the modified flux-ratio equation (see appendix A) (Hodgkin & Keynes, 1955; Kedem & Essig, 1965):

$$\ln(\phi_{joc}/\phi_{jco}) = n[\ln(c_0/c_c) - z_j F \psi_{co}/RT] \qquad (7.9)$$

where the value of n is about 2.5. This indicates self-interaction of potassium fluxes – of the type which was found also for the squid axon.

A model that might behave this way is one that has ions passing in file through long, narrow pores, but electro-osmosis in wider channels may also be a possible cause (Dainty, Croghan & Fensom, 1963; Thain, 1973). In part, then, the well-known discrepancy between 'flux conductance' and electric conductance is due to positive self-interaction of potassium ions in the plasmalemma. This self-interaction means that we should not use equations (6.1), (7.6) or (7.7). When an empirical approach is used, a discrepancy between $\sum_j g_j$ and g_M persists, but we would expect it to be reduced if experiments could avoid troubles with the cell wall.

Hydrogen ions

Taking the view that the discrepancy between g_M and $\sum_j g_j$ was due to neglect of the partial conductance of hydrogen ions in the membrane, Kitasato (1968) studied *Nitella clavata* by voltage clamping. He did not observe much change in the efflux of K^+ when the membrane potential was varied through changes in pH_0. The response of the p.d. to pH was one of those phenomena that most people had carelessly attributed to the cell wall. When the cells of *N. clavata* were voltage-clamped at the estimated value of ψ_K, the clamp current changed when pH_0 was changed. This current, depending on pH, was interpreted as a proton flux, and the apparent g_H could be estimated from $J_H/(\psi_H - \psi_M)$. The conductance so calculated was quite close to g_M, and g_K was quite small.

In fact, the efflux of K^+ was probably underestimated since 2-min collection times were used and the diffusion delay in the cell wall may have been of this order. Notwithstanding this and a discrepancy in the voltage-clamp records (fig. 11 of the reference, where the cell seems to have been much depolarised compared with the others), it appears difficult to explain

99

these results without recourse to the idea of proton conduction. Whether all or much of the effect of pH on p.d. is due to this is doubtful. For example, Lannoye, Tarr & Dainty (1970) and Richards & Hope (1974) found that lowered pH caused an increase in P_{Cl} and concluded that the depolarisation between pH 5 and pH 4 was substantially due to this factor. At the other end of the pH range (8–9) the p.d. is much hyper-polarised, more negative than ψ_K, but rather variably so in various charophytes, as shown in fig. 7.7. Kitasato (1968) concluded that the p.d. was hyperpolarised by the action of an outwardly-directed, active proton transport system, and that the action of this was to produce a steady pH in the cytoplasm by balancing a passive influx of protons.

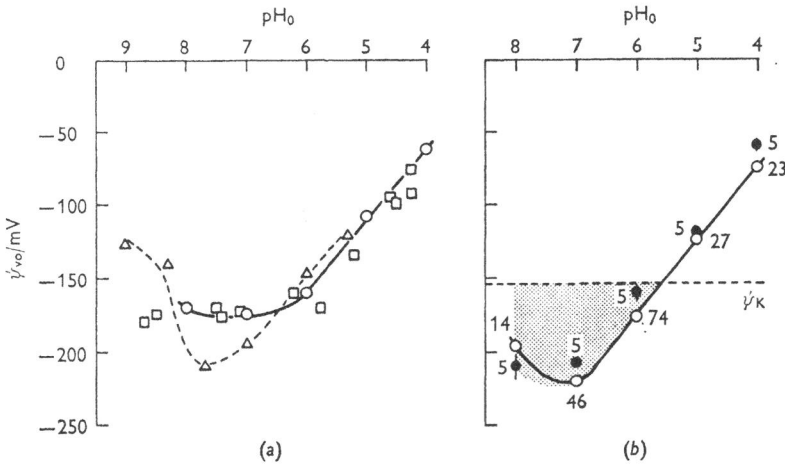

Fig. 7.7 (a) Effect of pH of the medium on ψ_{vo} in *N. clavata* (○, replotted, from Kitasato, 1968), *N. mucronata* (△, from Adrianov, Vorobieva & Kurella, 1968) and *N. translucens* (□, from Lannoye, Tarr & Dainty, 1970). (b) As for (a), for *C. corallina* in light (○) and dark(●) with number of cells used to calculate mean and SEM indicated. The significance of the shaded portion is discussed in the text (from Richards & Hope, 1974).

The present interpretation of relationships such as in fig. 7.7 is thus quite complex:

(i) In the shaded region of fig. 7.7b, ψ_M has an electrogenic component presumably due to an active H^+ efflux. At high pH, this pump may slacken off.

(ii) The change in ψ_M with pH_0 is partly due to the electrogenic proton pump being itself dependent on $[H^+]_0$, partly due to the importance of the term $P_H[H^+]_o$ in a more general equation than (7.1) for ψ_{1o}, and partly due to changes in P_{Cl}.

(iii) The loss of response of ψ_M to changes in $[K^+]_o$ in solutions of pH > 6 (approx.) presumably means that $P_H > P_K, P_{Na}$.

The fact that in the unbuffered solutions commonly used (pH $\approx$ 5.5), ψ_M is about equal to or more positive than ψ_K, and is often responsive to changes in $[K^+]_o$ obscured for years any contribution made at pH 5.5 to the conductance by the electrogenically produced H^+ fluxes. Details of the evidence for the above interpretation of the effects of H^+ on the p.d. is given by Richards & Hope (1974).

The pH of the medium has less effect on membrane conductance than might perhaps be expected in a system where most of the passive flux is supposed to be due to protons. Even if much of the membrane conductance is vested in the electrogenic pump itself (see Spanswick, 1973) the rate of the latter should vary with pH.

Effects of pH on the p.d. in giant cells other than charophytes have not been systematically studied. In *Griffithsia*, hyperpolarisation of the plasmalemma is not observed and $\partial\psi/\partial[H^+]_o$ is very small.

Effects of light

Light has some intriguing effects on p.d. and resistance that defy analysis so far. Table 7.2 summarises some of these with explanations where these were suggested. These varied results, mostly not interpreted, are rather confusing. The primary physical event is often not identified. Since light shone on previously darkened cells starts a number of known processes, the rational explanation of the primary effect should begin with these. We know also (chapters 9 and 10) that there are light-stimulated chloride and potassium transports at the plasmalemma. Also, fixation of carbon dioxide may alter HCO_3^- concentrations in various locations within and without the cell.

Short-term effects of light, and photosynthesis

The effect of light on membrane p.d. and resistance in *N. translucens* has recently been studied by Vredenberg (1969, 1970a, b) but interpretation still remains a problem. His

TABLE 7.2 *Effects of light on membrane p.d. and conductance*

	Solution (mM)	Effect on ψ	Effect on g	Comment	Reference
N. flexilis	0.1 KCl or NaCl	Hyperpol. −143 D −171 L	—	pH? Long-lasting L effect	Nagai & Tazawa (1962)
	0.9 KCl + other salts	Hyperpol. −100 D −138 L	—	pH ~ 5 Temporary L effect	Volkov (1964)
	0.1 KCl	Depol. −158 D −135 L	—		Adrianov, Kurella & Litvin (1965)
C. braunii	0.05 KNO$_3$, etc.	Hyperpol. −260 D −305 L	—	Light-adapted cells	Hansen (1971) Nishizaki (1963)
	0.05 KNO$_3$, etc.	Hyperpol. −175 D −205 L	—	Highest recorded p.d.? 24 hour predarkened	
	0.5 KCl 0.5 CaCl$_2$ 0.2 NaCl	Hyperpol. −136 D −190 L	Increased in L	24 hour predarkened	Nishizaki (1968)
	0.5 KCl 0.5 CaCl$_2$ 0.2 NaHCO$_3$	Depol. −165 D −135 L	—		

C. corallina	APW	−	−	L effects not observable in all-chloride media	Hope (1965)
	APW+0.1–0.5 HCO_3	Increased in light 0.33 D 1.4 L	Hyperpol. −165 D −215 L	Long-lasting effect (>1–2 hours)	Findlay et al. (1969)
	FPW	No consistent change	Small hyperpol. −131 D −137 L		
N. translucens	APW	Decreased 0.30 D 0.37 L	Depol. −135 D −131 L	Interpreted as P_{Na} reduced in dark, P_K unchanged	Hogg, Williams & Johnston (1969)
H. reticulatum		−	Hyperpol. −85 D −140 L		Metlicka & Rybova (1967)
Acetabularia crenulata	SW	−	Hyperpol. $\Delta\psi \sim$ 40 mV	Action spectrum showed λ700 nm quite effective. Light absorbed by ps pigments responsible	Schilde (1966)
Acetabularia mediterranea	ASW	−	Hyperpol.	P.d. sensitive to $\Delta[K^+]_0$ both in L and D. Hyperpol. often persisted in D. Hyperpol. assoc. with large light-stim. Cl^- influx	Saddler (1970)

D: dark conditions; L: light conditions, intensity often not specified; ps: photosynthesis.

observations were during the first few minutes of light. Under these conditions light caused a depolarisation of $1-4$ mV, light of wavelength 705 nm being 88% as efficient as that of 663 nm. That is, the phenomenon of 'red drop' was absent, as it was too with *Acetabularia* (Schilde, 1966). However 1 μM DCMU abolished the light-induced change in ψ_{vo}, suggesting PS II is crucial, and this contradicts the evidence regarding relative quantum efficiency in far red light. This is resolvable (if we retain the series scheme, fig. 9.1) by accepting Vredenberg's suggestion that the absorption was underestimated at 705 nm (personal communication).

Vredenberg identifies the primary event as a light-induced membrane current. The quantum requirement of this was measured in other experiments by noting the current required to hold ψ_{vo} at the dark level when light at 435 nm was switched on. It was estimated that $5-15$ quanta per ion were needed. The lower quantum requirements were obtained when the APW medium contained 1 mM Ca^{2+}, rather than 0.1 mM.

Vredenberg proposes that a light-induced proton uptake from cytoplasm to chloroplasts leads to changes in cytoplasmic $[H^+]_c$ that are reflected in the p.d. across the plasmalemma. $[H^+]_c$ and P_H may well be factors affecting the p.d. under certain conditions, especially with 1 mM Ca^{2+} in the medium. However, indications at present are that the proton shift observed with isolated broken (class II) chloroplasts is not obtained with intact (class I) chloroplasts (Heber & Krause, 1971) perhaps due to proton impermeability of the chloroplast envelope. Thus, *in vivo*, the occurrence of a proton shift between the chloroplasts and the surrounding cytoplasm would appear doubtful.

In a later report (1973) Vredenberg has attempted to interpret changes in the current:voltage characteristic of the *Nitella* membrane under various conditions of light and pH. He does this in terms of changes of the current in parallel with, and the voltage in series with, an invariant nonlinear resistance. His method then involves arbitrarily moving the observed voltage:current curves along both axes until they coincide. The distance moved is said to give the extra current and voltage from a membrane generator. His own results, however, show that the current:voltage characteristic changes shape as well, which undermines the basis of the method.

The lines of work opened up by Vredenberg seem well worth pursuing.

Long-term effects of light on p.d. and conductance

Spanswick (1973) has concentrated on the longer-lasting effects of light on the p.d. in $N.$ translucens, where for some hours a hyperpolarisation is observed using a solution of pH 6 and containing 0.4 mM K^+. Rather striking changes in conductance between light (0.59 S m^{-2}) and dark (0.065 S m^{-2}) were also noted, which were absent in $C.$ corallina unless there were bicarbonate present (Hope, 1965). In this phenomenon, a pH of 7 does not substitute for bicarbonate, but at pH 8, the conductance may decrease in the dark (Richards & Hope, 1974). Spanswick (1973) stresses that the putative proton pump may be light-stimulated and the conductance, being due to this pump, would be therefore much decreased in the dark, as observed. In $C.$ corallina the hyperpolarisation of the plasmalemma, when it occurs at $pH_0 < 7$, is not dependent on light; to this extent Chara is consistent with Nitella, in a twisted way, because its conductance is not light-sensitive. Placing the onus on the active transport system to supply the missing conductance (i.e. that not supplied by K^+, Na^+, Cl^-) is a stimulating alternative theoretical idea to that of 'non-independent' fluxes which we expounded earlier.

Response to sinusoidally-varied intensity

In a more general approach, Hansen & Gradmann (1971) and Hansen (1971) analysed the response of the p.d. to sinusoidally-modulated light, in Acetabularia and $N.$ flexilis respectively. Interesting resonance effects and phase shifts were found and considered in terms of network theory and feedback systems.

Permeability modifiers

In view of the spectacular effects of substances such as the macrocyclic depsipeptides, (valinomycin, the gramicidins) and polyene antibiotics (nystatin, amphotericin) on artificial lipid bilayers, their effects were tried on membranes in Chara (Findlay, Hope & Sydenham unpublished data).

Valinomycin acts by forming readily a complex with potassium (Pressman, Harris, Jagger & Johnson, 1967), which has a far greater solubility in the membrane than K^+ alone. Both the complex and the valinomycin molecule are mobile in the membrane. Hence the bilayer becomes K^+-permeable on addition of valinomycin and its resistance decreases greatly in the presence of K^+ (Andreoli, Tieffenberg & Tosteson, 1967). Very little effect of up to 2.10^{-5} M valinomycin was noted in *Chara* cells. However Cram (personal communication) has found that ethanol or other solvents used in initially dissolving the valinomycin must be rigidly excluded from the experimental solution. If this is done an effect on p.d. is obtained; the cell resistance was not monitored in that experiment. Since the algae studied are already rather K^+-selective in the plasmalemma, valinomycin may have a less noticeable effect than on ultrathin lipid layers.

Nystatin, on the other hand, makes artificial membranes anion-selective and increases their conductance by orders of magnitude, possibly by forming conducting pores within the membrane (Finkelstein & Cass, 1968). In *Chara*, nystatin depolarises and eventually at 5×10^{-5} gm ml^{-1} renders the plasmalemma inexcitable (see chapter 8) and decreases its resting resistance 10–30 fold. The ionic fluxes that flow in response to the depolarisation and resistance decrease are consistent with a huge increase in chloride permeability in the resting state. The details of the molecular interaction of nystatin with either natural or artificial membranes is still obscure.

Uncouplers such as CCCP and FCCP are members of another class of substances, which cause an increase in membrane permeability to protons. Indeed their uncoupling action in Mitchell's hypothesis (1966) comes about because a metabolically-maintained gradient of electrochemical potential for protons, said to drive phosphorylation, is destroyed by the induced proton permeability. It seems important to find whether uncoupling substances have generally an effect of inducing proton permeability in natural membranes, as they seem to in artificial ones.

Temperature

It is expected that short-term temperature changes may be made to algal cells without changing internal concentrations, while permeabilities and reaction rates should be directly affected. A study of the influence of temperature on membrane p.d. and conductance was made by Hogg, Williams & Johnston (1968b) in *Nitella* and by Hope & Aschberger (1970) in *Chara* and *Griffithsia*. Both groups interpreted these results in terms of temperature effects on ionic permeabilities. In these coencytes the authors felt that there was firm evidence for regarding the plasmalemma p.d. and conductance as being determined almost entirely by P_K, P_{Na}, $[K^+]_o$, $[K^+]_c$, $[Na^+]_o$ and $[Na^+]_c$. P_K and P_{Na} were calculated by both groups from observations of ψ_{co} and g_{co} at various temperatures, and from fluxes by Hope & Aschberger. It can be seen (fig. 7.8) that log P_K and

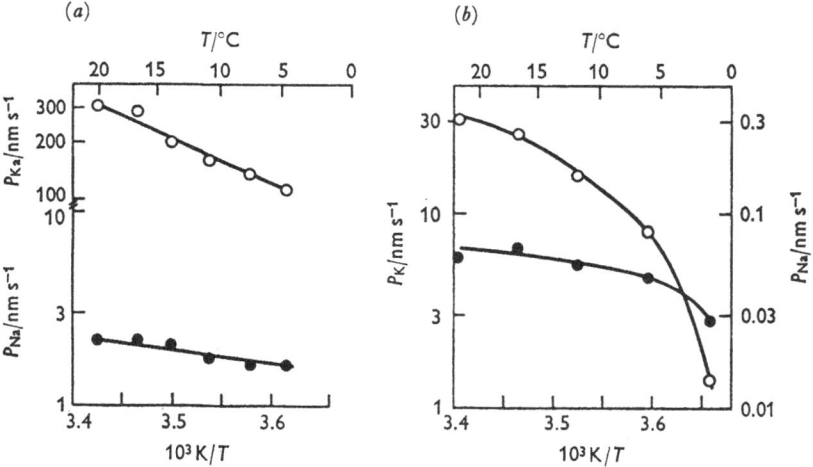

Fig. 7.8 (a) Permeabilities of the plasmalemma of *G. pulvinata* (P_K, ○; P_{Na}, ●), calculated from data on change in ψ_{co} and g_{co} with temperature, and plotted on a log scale against reciprocal of absolute temperature. (b) As for (a) but permeabilities calculated from influxes of K^+ or Na^+ at various temperatures (from Hope & Aschberger, 1970).

log P_{Na} are nearly linearly related to $1/T$, corresponding to a constant enthalpy of activation. In the absence of other factors this would be the activation energy for an ion to penetrate the membrane. The slopes of these relationships (a rather unsatisfactory average has to be taken for P_K obtained from flux data) are about the same for both electrical and flux measurements. However, the permeabilities differ greatly in the two methods of estimation, and this in a cell that apparently has no electrogenic pump to resolve the discrepancy. The activation enthalpy is greater for K^+ than Na^+, which is significant, because P_K is so much greater than P_{Na}. No large difference of this sort exists for permeation through cation exchange resin (Hope & Huber, unpublished data) so it is concluded that the sodium permeation is in the nature of a leak, with a few rather large pores allowing Na^+ to move with few frictional interactions with the membrane.

On the other hand K^+ permeation seems to be via a separate pathway where there are significant energy barriers to be overcome. A small amount of K^+ would be expected to permeate the Na^+ pathway, as well. These conclusions, especially those from experiments with *Chara*, must remain provisional because of the possible effects of temperature on the proton transport system as well as on permeabilities.

Electro-osmosis

Electro-osmosis is the flow of water across a barrier, produced by electrical driving forces. Where it occurs, for example in ion-exchange resins, it is thought of as resulting from interaction between ions migrating in the electric field and water molecules in the same paths of migration. In natural membranes, one would infer from the interaction that ions and water follow common paths through the membrane, which is an open question before the finding of such an interaction. We have seen in chapter 3 that attempts to prove or disprove the existence of pores, through which water moves in osmosis, have not in the main been successful. However in the study made by Gutknecht (1967*b*) with *Valonia*, all water movement can be accounted for by independent diffusion. The extent of electro-osmotic coupling has been calculated in certain instances from models of the interaction by Dainty, Croghan & Fensom (1963). They used the frictional model where interaction is formally equivalent to a frictional force between water and ions, proportional to the frictional coefficient and to the mean relative velocity between the ions and water molecules (Spiegler, 1958). The inverse electrokinetic phenomenon of the streaming potential is the p.d. which results from an imposed water flow through the membrane. It is a result of the same coupling between ions and water and its occurrence (for example in gall bladder – Pidot & Diamond, 1964) allows the same inferences about water-filled, electrically charged regions for ion conduction.

Fensom & Dainty (1963) first observed electro-osmosis in cells of *N. translucens*. The experiments are difficult to perform, since it is necessary to detect water flow rates of the order of 10^{-9} to 10^{-10} l s^{-1}, if the current density is to be kept below 0.1 A m^{-2}. In these experiments the water flow was detected in much the same way as it was for the transcellular osmosis work (chapter 3). It was observed that the rate of water flow increased for many minutes after the current was applied. It is to be expected that the electro-osmotic flow would begin at once and remain constant. As subsequently shown for *Chara* by Barry (1967), when he developed equipment able to detect small flows within seconds, there is an immediate flow of 7 μl C^{-1}, with a slow increase to a maximum of 18 μl C^{-1}. The prompt flow is identified as electro-osmotic; the later increase is due to osmosis through a membrane (probably the plasma-lemma) across which is developed a small osmotic gradient during the passage of current. An account of the theory and experiments relating to local osmosis and electro-osmosis both in cells and wall segments, is given by Barry & Hope (1969a, b). A further complication is that the cell wall also exhibits electrokinetic effects and these may partly contribute to current-induced water flow in cells (Briggs, 1967; Tyree, 1968).

It is possible that up to 30 water molecules move across the plasmalemma with each equivalent monovalent cation, a figure arrived at by Barry & Hope (1969c) after attempting to allow for the cell wall effects. Tyree & Spanner (1969) also concluded that the electro-osmosis shown by living *Nitella* cells is unlikely to be due solely to the cell wall. We can endorse ruefully the remarks of MacRobbie & Fensom (1969) about measurements of electro-osmosis in *Nitella*, '... even in this, the simplest electro-osmotic system studied, the situation is a good deal more complex than has been assumed and some caution is needed ...'.

Our tentative conclusion is that there is electro-osmotic coupling between water and cations in the plasmalemma, which suggests that cations penetrate the membrane through 'pores' where they are associated with water.

The capacitance of membranes

As well as acting as conductors of electric current, membranes exhibit a capacitance, which has usually been attributed to the presence of a non-conducting layer of lipids of low dielectric constant. An early thumbnail calculation assumed a capacitance of about 0.01 F m^{-2} and a permittivity of 2.7×10^{-11} F m^{-1} (this corresponds to a dielectric constant of 3, as in olive oil), and yielded the 'thickness of the lipid membrane' as 2.7 nm.

In *Chara* (Findlay & Hope, 1964) and in *Nitella* (Williams, Johnstone & Dainty, 1964) the value of the membrane capacitance is 0.01 to 0.02 F m^{-2}. Earlier studies (Curtis & Cole, 1937; Cole & Curtis, 1938) had shown the capacitance to be of this value and invariant during the passage of an action potential: studies of the frequency dependence of the impedance of *Nitella* showed the capacitance to be 'lossy', with a constant phase angle a little less than $\pi/2$ rad.

There the matter rested until Coster & Smith (unpublished results – see Coster, 1973*a*) examined the impedance of *C. corallina* over a wide range of frequencies, including very low frequencies where electrode impedance used to bedevil workers. The technique used was to pass alternating current between an inserted, longitudinal, metal electrode (Findlay & Hope, 1964) and an external spiral electrode, with two other electrodes to measure changes in p.d. across the membrane. The capacitance was also examined as a function of membrane potential. The results are interpreted as arising from a membrane which is essentially a junction of regions of opposite fixed charges – the same model which predicts the 'punchthrough' effect. This interesting challenge to the established dogma grows steadily stronger.

We see in retrospect that electrical measurements, originally undertaken in the expectation that plant cells might follow some of the rules proposed for animal cells such as nerve and muscle, have led in their own right to critical re-examinations of models for membrane structure, and of the ramifications of electrogenic pumps.

CHAPTER 8

Action potentials in charophyte cells

Introduction

One reason the charophytes are famous among a wider audience than botanists is that the cells can respond to stimulation by conducting an electric depolarisation along their length. Perhaps if the phenomenon in plant cells had not been so infernally leisurely, taking several seconds to complete, the inner nature of the action potential would have been revealed earlier than it was. This is not to say that the molecular events accompanying the action potential in either plants or animals have been worked out, but at least a description has been obtained in ionic and electrical terms.

Unfortunately it is not possible to include with the book a moving film of cytoplasmic streaming and its cessation during an action potential; these are worth observing should the opportunity arise (see chapter 12).

No giant algal cells other than the charophytes seem to be excitable, though Umrath (1938) reported long-lasting electrical responses to strong shocks in *Valonia* cells. These responses do not seem to have been further studied. Also a regenerative depolarisation of the *Acetabularia* membrane has been termed by Gradmann (1970) an action potential; transmission along the cell has not yet been reported.

Features of the action potential

Its similarity to the action potential in animal cells includes the following aspects –
  (a) Stimuli such as a depolarising current, suddenly lowered temperature, and certain chemicals are effective.
  (b) When the plasmalemma is depolarised past a sharp threshold level, the AP always occurs, in the form of a further large depolarisation and recovery.
  (c) The peak depolarisation does not usually depend on the strength of the stimulus (all-or-none response).

(d) There is a refractory period after an AP during which no stimulus can cause a second AP.

(e) The AP is conducted in both directions away from the place at which the plasmalemma is first taken past threshold.

(f) The 'strength-duration curve' for a sufficient stimulus is roughly hyperbolic.

Figure 8.1 depicts an action potential in *C. corallina*, recorded from the vacuole. The fast and slow components marked A and B were revealed to be separate action potentials across the plasmalemma and tonoplast, when appropriate insertions and recordings were made. Fig. 8.2 shows these separate APs; it is seen that the sum of ψ_{co} and ψ_{vc} at any time yields the composite AP of fig. 8.1.

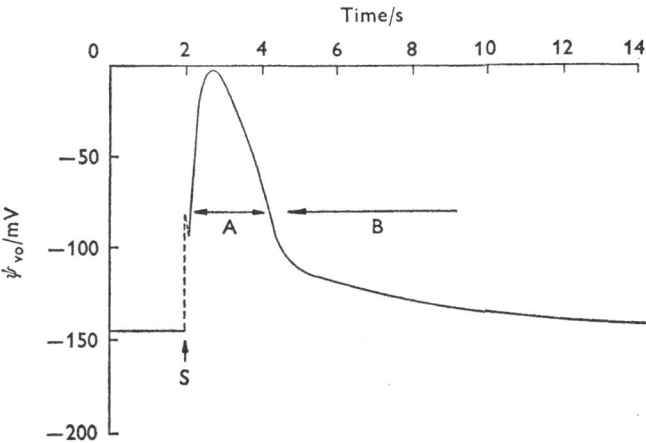

Fig. 8.1 Continuous recording of ψ_{vo} in a cell of *C. corallina* showing an action potential following a depolarising stimulus at $t = 2$. A and B are 'fast' and 'slow' components referred to in the text: S, stimulus.

The two action potentials almost always occur together, perhaps because that at the plasmalemma stimulates the tonoplast. The antibiotic nystatin (see chapter 7) may abolish the plasmalemma AP while leaving the tonoplast excitable, but it has not been possible to fire the plasmalemma alone.

The conductance of the membrane during the AP had earlier been shown to increase (Cole & Curtis, 1938), reaching a maximum at the peak and then returning to normal. In fact both plasmalemma and tonoplast behave this way, as shown in fig. 8.2, the plasmalemma having relatively the greater increase in conductance.

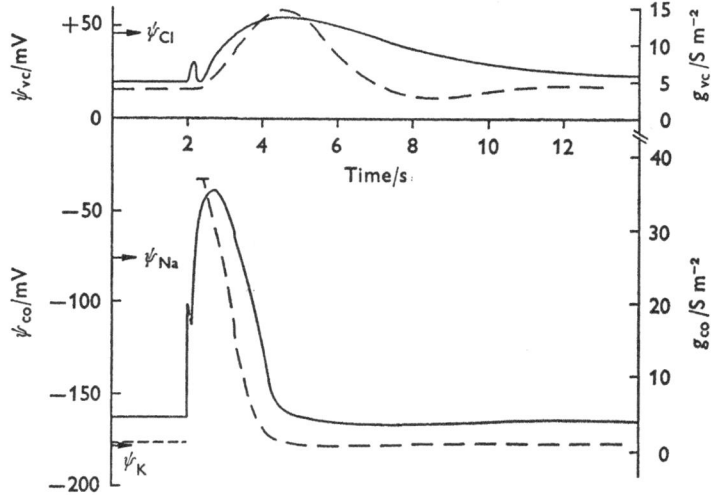

Fig. 8.2 Recordings of ψ_{co} and ψ_{vc} in the same cell as in fig. 8.1 made a few minutes after the action potential shown there. The dashed lines are the approximate time courses of the conductances of the plasmalemma and tonoplast during the respective action potentials, based on results of Findlay & Hope (1964). ψ_K, ψ_{Na} and ψ_{Cl} are estimated Nernst potentials for the plasmalemma for these cells.

Factors affecting the AP

It is known that stimulation produces in nerve axons a rapid increase in sodium permeability, and that as a consequence the p.d. changes from near the equilibrium potential for K^+ (about -80 mV, the resting state) towards that for Na^+, which is in the neighbourhood of $+40$ mV. The spontaneous decrease in sodium permeability and the increase in potassium permeability produced by depolarisation return the membrane p.d. to its resting level within a few milliseconds. The nerve AP is possible because the sodium gradient is in the opposite direction to the potassium gradient. The peak of the AP does not reach ψ_{Na} because the inrush of Na^+, which makes the inside positive, is eventually balanced by the rising efflux of K^+. The efflux of K^+ rises because the p.d. moves further from ψ_K, and because P_K increases. The peak occurs when the currents balance, somewhere between ψ_K and ψ_{Na}, but much closer to ψ_{Na}.

With this explanation in mind, experimenters began to try the effect of various ions in the medium upon the action potential in plant cells, and to measure the extra fluxes of ions accompanying the AP. The rationale for the former experiments lay in the expectation that the peak of the AP should depend on the concentration of the ion species to which the plasmalemma became transiently permeable. This peak may approach the equilibrium potential for that ion, and clearly should be a function of its external concentration. The equilibrium p.d.s for K^+ and Cl^- can be calculated for *Chara* cells in APW (table 5.1) from the data for the concentrations of these ions in the cytoplasm (table 5.5). The sodium concentration in the cytoplasm could be taken as somewhere near 20 mM (Kishimoto & Tazawa, 1965a). Choosing the 10 mM result for chloride, we obtain $\psi_K = -178$ mV, $\psi_{Na} = -76$ mV and $\psi_{Cl} = +46$ mV. A guess at ψ_{Ca} puts it at about zero or positive, if the cytoplasm activity is of the order of 0.25 mM or less. These Nernst potentials are shown on the diagram for the plasmalemma action potential in fig. 8.2. It is seen that the ions which might produce the AP are calcium and chloride.

The chloride hypothesis

Gaffey & Mullins (1958) did experiments with *Chara globularis* in which they detected an increased efflux of K^+ and Cl^- in stimulated cells, and in which Cl^- in the medium was shown to be essential to maintain excitability. Further, added choline chloride affected the peak of the AP in the way expected: that is, increased $[Cl^-]_0$ caused the peak to occur at a more negative potential. They therefore suggested that a transient increase in Cl^- permeability caused the AP.

Findlay (1961, 1962) and Hope (1961) were not inclined to accept the role of chloride ions because of the marked effect of $[Ca^{2+}]_0$ on the peak of the AP and on transient, voltage-clamp currents in *C. corallina* (fig. 8.3). This effect, apparently not found in *Nitella axillaris* (Kishimoto, 1964) but present in *Nitellopsis obtusa* (Findlay, 1970), caused them to conclude that the AP was produced by a transient Ca^{2+} influx.

From then on, evidence in favour of chloride accumulated. Mullins (1962) did more appropriate Cl$^-$ tracer measurements on *N. clavata* (an ecorticate species, unlike *C. globularis*), Hope & Findlay (1964) did tracer studies with Ca^{2+} and Cl$^-$ on *C. corallina*, and Kishimoto (1964) studied voltage-clamp currents in *N. axillaris*. All three studies showed that chloride, not calcium, produced the AP. Hope & Findlay found that the

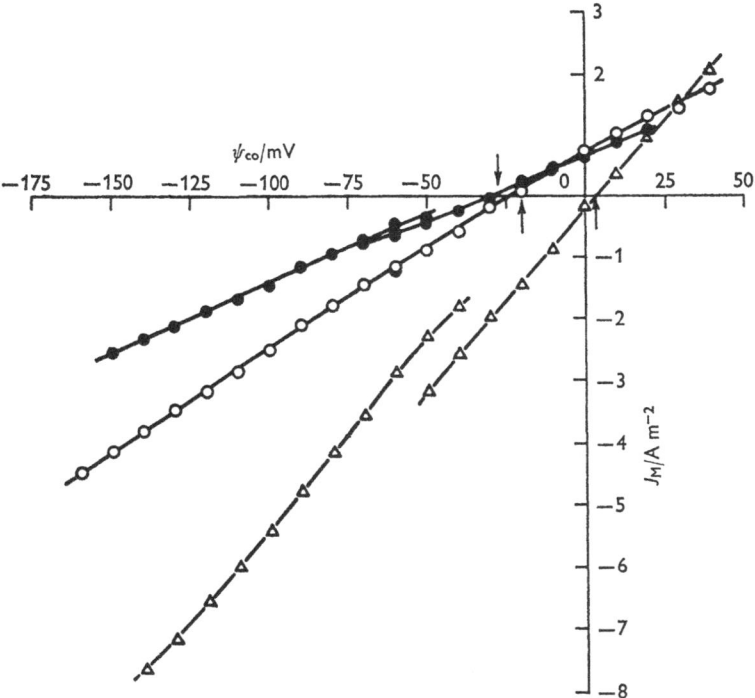

Fig. 8.3 Voltage-clamp currents at the peak of excitability in the plasmalemma of a cell of *C. corallina* plotted against ψ_{co}, the clamped p.d. Values of $[Ca^{2+}]_0$ were 0.15 mM ($\bullet$), 0.5 mM ($\circ$) and 1.5 mM ($\triangle$). The scans were obtained with the aid of a ramp generator (from Findlay, 1964).

extra chloride efflux accounted for the charge transfer across the membrane in their voltage-clamp studies. In both *Chara* and *Nitella* a single AP releases 5–10 μmol m^{-2} of Cl$^-$. An extra efflux of K$^+$ has been shown, but has been less quantitatively studied. The chloride permeability undergoes a transient rise to some 10^{-6} m s^{-1}, at about 0.5 s after stimulation, and has fallen to a low level again by 2–3 s. In the excited state, the

E

membrane of *Chara* does not distinguish between chloride, bromide and nitrate, since clamp currents are similar as interchanges between these anions are made in the medium, the total concentration being 20.1 mM. There is not much evidence for changes in P_K, the extra efflux of K$^+$ being perhaps produced by the change in ψ_{co}. The Cl$^-$ and K$^+$ ions leaving the cell diffuse with some delay through the cell wall – a delay which has been shown directly by Mailman & Mullins (1966) for Cl$^-$. The ion movement is accompanied by a water movement, so that there is a momentary loss of pressure of 1–3 kPa (Barry, 1970). This is $<0.3\%$ of the TP. Whether this is an osmotic effect due to the diffusion delay, or an electro-osmotic effect, is now to be considered.

Models for the action potential

Empirical descriptions of the action potential in animal cells, for example that of Hodgkin & Huxley (1952), fit equations to the changes in p.d. and conductance, with the aid of a number of rate constants and other parameters. This approach has not been carried over into the world of plant action potentials, mainly because the data for conductance changes ('activation' and 'inactivation') are not available. Even without this description, it may be possible to consider the applicability of some physical models; such a one is Teorell's (1959). It is based on electrokinetic phenomena in large pores, and predicts that flow of water should accompany ion flow during the action potential. This is testable in plant cells where water flows produce just measurable pressure changes. One has the feeling that it would not be possible to extract information from giant axons about internal pressure changes taking place in milliseconds.

Barry (1970) indeed observed that small volume flows accompany the AP in *Chara*, the maximum rates of pressure change and of volume flow occurring close to the time of the peak of the action potential. The lack of an appreciable delay suggests that not much of the flow is due to local osmosis. The magnitude is explained by electrokinetic coupling with both the chloride and potassium effluxes. Barry states that the volume flows and pressure changes may have been incidental consequences of the changes in membrane permeability, and

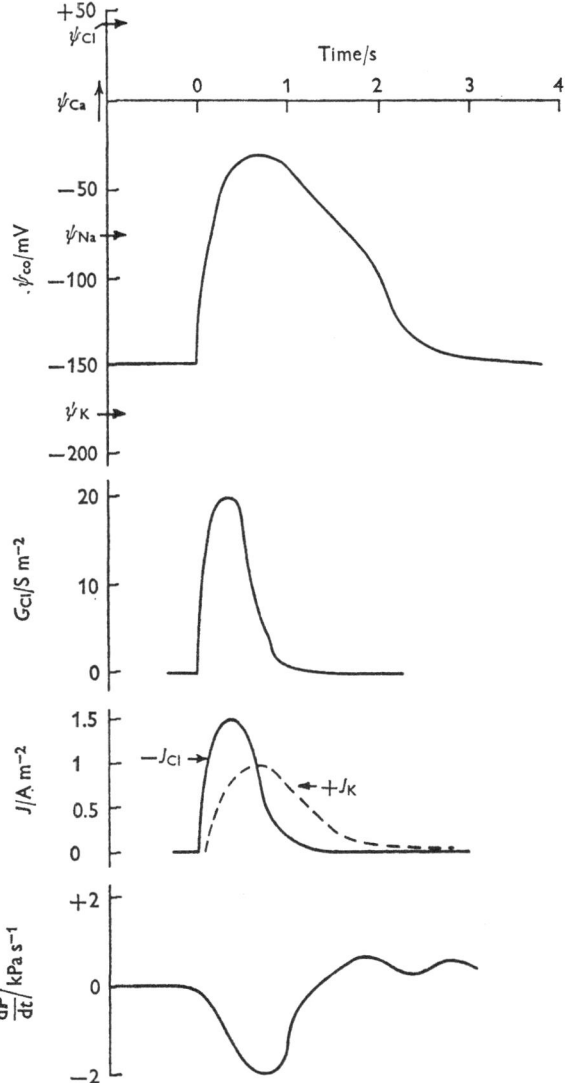

Fig. 8.4 The time-sequence of postulated events during an action potential across the plasmalemma after a supra-threshold stimulus at $t = 0$. Approximate scales for g_{Cl} and J have been established from values of peak conductance during an AP and from integrated effluxes of Cl^- (tracer experiments). dP/dt after Barry (1970).

117

though he considers the Teorell model at some length, rejects it on *a priori* grounds rather than by showing it to be inconsistent with the data.

We can thus give some description of the course of events during a (membrane) action potential – for greater simplicity we are assuming the membrane to be stimulated all at once with no propagation. The onset of the AP is marked by a sharp increase in g_{Cl}, which gives rise to a flow of inward current carried by a chloride efflux. The depolarisation produced by this current increases g_{Cl} still further; it then begins spontaneously to decrease again: meanwhile an outward current carried by K^+ efflux has been rising rapidly as ψ rises. When these currents are equal the peak depolarisation has been reached: thereafter g_{Cl} continues to decline, ψ becomes more negative and J_K becomes smaller too as a result. Eventually ψ is back at its resting level, but for a short time the inactivation of g_{Cl} remains, and the membrane is refractory. Both J_{Cl} and J_K sweep water out of the cell, giving a transient fall in pressure. Fig. 8.4 shows the likely time course of some of the observed and calculated quantities.

A similar mechanism of a transient increase in the permeability to chloride has been proposed for the tonoplast (Findlay & Hope, 1964). The direction in which the p.d. moves is consistent with this, because ψ_{Cl} is about +60 mV, which is approached during the AP.

Changes in the cytoplasm

Observations of the chloride activity in the cytoplasm during an action potential (Coster, 1964) showed that this activity decreased by several mM during the first 1–2 seconds after the rising phase of the plasmalemma AP. The chloride activity then returned to the resting level in the next 6–8 seconds. This latter time corresponds to the duration of the tonoplast AP. Thus the chloride in the cytoplasm seems to be replenished by a large, transient, passive flux from the vacuole. The initial decrease in chloride activity of 2–3 mM corresponds to an efflux of about 12 μmol m^{-2} from a phase 5 μm thick, which is not far from the extra efflux usually observed for one AP.

When two electrodes, one filled with 0.2 M KCl, the other 2 M KCl, were placed in the cytoplasm of *Chara*, a steady p.d.

of about 40 mV was observed between them, the 0.2 M electrode being positive (Coster, Syriatowicz & Vorobiev, 1968). This is consistent with the cytoplasm being a cation-exchange phase, with negative fixed charges in the resting condition. However, Walker (1955) observed a much smaller p.d. in similar experiments. Neither set of data can be used to calculate the fixed-charge concentration without knowledge of ionic mobilities.

In the experiments of Coster et al., the p.d. between the two KCl-filled electrodes reversed in sign during the AP, so that the probe with 0.2 M became −30 mV with respect to the other, returning to +30 to 40 mV on completion of the AP. One explanation of this is that the net charge on the fixed ions changed sign from negative to positive, and back, during the AP. This could happen through a transient appearance of protons in the cytoplasm, but there is no evidence for this at present.

Conduction of the action potential

The action potential is conducted away from the spot initially fired by a flow of local currents that are sufficient to bring adjacent membrane areas above threshold. Because of the necessity of local external currents, conduction can be slowed and eventually blocked as the resistance in the external solution rises. The local administration of anaesthetics or of KCl (which depolarise the plasmalemma and provide a current sink) also block conduction, an old observation. The conduction rate, when cells are in media comparable in conductivity with APW, is of the order of 0.01 m s^{-1} (Umrath, 1930).

Under some conditions an action potential initiated in one cell may be conducted across a node to the adjacent cell (Sibaoka, 1966; Spanswick & Costerton, 1967). This is consistent with a fairly low resistance pathway joining the cells, through plasmodesmata in the nodal region (plate 5).

CHAPTER 9

Studies of active transport–cations

The nature of active transport

In beginning to speak of 'active transport' it is difficult to achieve accuracy and avoid pedantry, difficult to define active transport accurately and to conform to biological usage.

Uphill transport is easy to define, and relatively easy to distinguish experimentally. Transport of a substance across a membrane may be called 'uphill' if there is a net movement from a lower to a higher electrochemical potential – this corresponds to a net increase in the free energy of the substance. The inference may be drawn that there is a transfer of energy between some system and the substance in question – there is an interaction in the membrane.

The possible interactions in membranes are formalised in the equations of steady-state thermodynamics: Kedem & Essig (1965) have for example provided a version of the flux-ratio equation (Ussing 1949, Teorell 1949) which may serve to illustrate the formal possibilities:

$$RT \ln(\phi_{jab}/\phi_{jba}) = (\mathscr{R}^*/\mathscr{R})(\bar{\mu}_a - \bar{\mu}_b) + \sum_{k=1}^{k=m} \mathscr{J}_k \int_0^\delta (r_{jk}/r_{jj}) \, \mathrm{d}x$$

$$+ \mathscr{J}_n \int_0^\delta (r_{jn}/r_{jj}) \, \mathrm{d}x \qquad (9.1)$$

(see appendix A).

In this equation the ratio of the unidirectional fluxes is determined by three terms. The first is the difference in electrochemical potential of the species, with a factor $\mathscr{R}^*/\mathscr{R}$ which is one for the case of passive, independent diffusion, and may be $\lessgtr 1$ for the case of self-interaction of the fluxes of j. The factor $\mathscr{R}^*/\mathscr{R}$ has only positive values, so this term represents only downhill transport. The summation term represents the number of possible interactions between the flux of j and the fluxes of all other components of the system. Each represents an exchange of energy, and each in principle may have the sign opposite to that of $(\bar{\mu}_a - \bar{\mu}_b)$. Thus each term $\mathcal{J}_k \int_0^\delta (r_{jk}/r_{jj})\,\mathrm{d}x$ may tend to move j against its electrochemical potential gradient. Any one, or their sum, may exceed the first term in magnitude, and thus give a net flux in the 'uphill' direction. The last term in the equation represents the coupling of the flux of j to a scalar metabolic reaction whose rate is $\mathcal{J}_n$: it represents an exchange of energy between a metabolic reaction and a membrane flux such as would characterise the usual 'active pump' of the biologist. Here too the term may provide an uphill transport of j.

Stein (1967) has labelled processes that are described by the third term 'primary active transport', and those described by elements of the second term 'secondary active transport'. Clearly neither process necessarily produces an uphill flow, though the finding of an uphill flow necessarily implies the presence of a primary or a secondary active transport. For this reason perhaps the terminology used by experimenters tends often to be restricted to the term 'active transport' itself: one can immediately infer 'active' from 'uphill', and one frequently does not know whether the mechanism is primary or secondary. This is true of many of the transport systems discussed here. In the terminology of Mitchell (1967) we do not yet know in many cases whether we have on our hands a uniport $(\mathcal{J}_n)$ or a sym- or anti-port $(\mathcal{J}_k)$. The way onward seems to be to elucidate mechanisms.

Satisfying though equation (9.1) may be, it is cold comfort to the experimenter, who is always unable to measure the sizes of its numerous terms. It is comfortless too in its ignoring the problem of mechanism, providing the same formal terms to describe for example frictional and chemical coupling mechanisms.

In the imperfect real world, a number of criteria have been used, many of them, as will be seen, more than a little inexact:

Passive	Active
1. No specificity, or competition	Specificity, competition
2. Downhill movement	Uphill movement
3. No inhibition	Inhibition by metabolic inhibitors
4. No specific inhibition	Inhibition by specific transport inhibitors
5. Rate $\propto$ concentration	Saturation kinetics
6. Low Q_{10}	High Q_{10}

Of these criteria, (1) really distinguishes mechanisms with specific interaction from those without; such mechanisms may provide either active or passive movement. Rule (2) is conclusive only if uphill transport is found, and then only in a semantic sense. The rule is of course exact for ions only if the electrochemical potential difference is used – the error of using only the chemical potential difference is fortunately becoming rare. Rule (3) is not very successful in real life, since whole cells react to metabolic inhibitors in complex ways that are ill understood. Rule (4) has the same distinguishing power as rule (1), if a specific inhibitor is found which has no metabolic effects. The difficulty of finding such inhibitors is among those 'facts of life' neatly explained by the Mitchell theory (1966). Rule (5) distinguishes plain diffusion systems from mechanised ones, much as do rules (1) and (4). Rule (6) distinguishes large from small enthalpies of activation, which is of some interest but not related by any compelling argument to the terms in equation (9.1). High temperature coefficients may characterise passive diffusion (chapter 7, and see Zwolinski *et al.*, 1949). Consideration is not often given to the rather different effects of short-term and long-term temperature changes on whole cells.

Some experimenters have sought to avoid the problems of equation (9.1) by using the simplified form, a truncated version of Ussing's (see appendix A),

$$RT \ln(\phi_{\mathrm{Jab}}/\phi_{\mathrm{Jba}}) = \bar{\mu}_{\mathrm{a}} - \bar{\mu}_{\mathrm{b}} \qquad (9.2)$$

This equation is derived from Ussing's by the omission of solvent drag terms which correspond to some of the terms in J_k in equation (9.1). Deviations from the equation do not constitute evidence for active transport, as Ussing himself pointed out. Plant physiologists still tend to use (9.2) without any estimates of drag terms, and without considering the wide range of values which $\mathscr{R}^*/\mathscr{R}$ may take on. There is already evidence for positive self-interaction in natural and artificial membranes (Hodgkin & Keynes, 1955; Meares & Ussing, 1959; Walker & Hope, 1969) and negative self-interaction is often postulated. Between them these effects can account for passive flux-ratios of any values whatever, within the limitation that the net flux is in a downhill direction.

The advantage of giant algal cells is chiefly that one can more or less readily determine $\bar{\mu}_{\mathrm{a}}$ and $\bar{\mu}_{\mathrm{b}}$ and even control both, although only for K^+ in *Nitella* and *Chara* is there any estimate of the value of $\mathcal{R}^*/\mathcal{R}$. Since the value found was 2.4, the use of (9.2) to calculate flux-ratios is futile.

Apart from the use of the flux-ratio equation, two complementary approaches have been used: (a) the cells are kept in constant conditions until a steady-state of zero net flux can be assumed or shown: this corresponds to the left-hand side of (9.1) being zero. A non-zero value of the first term on the right-hand side then implies non-zero values of the second or third, and active transport can be proclaimed; and (b) the value of $\bar{\mu}_a - \bar{\mu}_b$ (for vacuolar sap and external medium) is set at zero by Ussing's short-circuit technique. A non-zero value of left hand side of (9.1) then as before implicates active transport. That this method also assumes a steady state is clear.

Dainty (1962) has discussed the use of these tests. Generally, for those tests in which a non-zero value of $\bar{\mu}_a - \bar{\mu}_b$ was sought, (a) above, workers have used the Nernst equation (7.3) as the test for equality of $\bar{\mu}_a$ with $\bar{\mu}_b$. The easiest way to do this is to calculate from ψ and c_0 the expected value of c_i. The observed value can then be directly compared with this and the direction of a significant difference indicates the direction of the putative active transport.

It has been argued (Coster & George, 1968) that frictional interactions between metabolite fluxes and ion fluxes may account for all or many observed cases of uphill transport. In the approach adopted here such phenomena would be called active transport, and the thesis of Coster & George would be that: (i) primary active transport probably does not exist; and (ii) that secondary active transport can be attributed to frictional coupling with metabolic fluxes. There are few plant systems for which this thesis can be conclusively rejected. Most investigators however have found it too improbable a model to adopt, perhaps because the major metabolite fluxes in plant cells are those of carbon dioxide and oxygen – unlikely candidates for frictional interaction with ions, since one would expect them to penetrate lipid membranes readily by solution in the lipid, and not to go through polar ion channels.

Keynes (1969) has catalogued many features of active transport in numerous animal epithelia. He puts the minimum number of necessary ion pumps at five. There are:

(i) a sodium/potassium pump; sodium is transported out,

E*

TABLE 9.1 *Active transports inferred*

Material	Ions [a]			Type of test [b]	Reference
	K^+	Na^+	Cl^-		
C. globularis	Q	vo	ov	N	Gaffey & Mullins (1958)
C. corallina	Q	vo	–	N	Hope & Walker (1960)
N. translucens	ov	vo	ov	N	MacRobbie (1962)
N. translucens	($H_2PO_4^-$ ov)		–	N	Smith (1966)
H. africanum	ov	vo	ov	N, FR	Raven (1967)
Tolypella intricata	Q	–	ov	N	Smith (1968a)
Nitellopsis obtusa	Q	vo	ov	N	MacRobbie & Dainty (1958)
Lamprothamnium succinctum	–	vo	ov	N	Kishimoto & Tazawa (1965b)
Halicystis ovalis	Q	vo	ov	SCP	Blount & Levedahl (1960)
Chaetomorpha darwinii	ov	vo	Q	N	Dodd et al. (1966)
Valonia ventricosa	ov	vo	Q	FR	Gutknecht (1966)
V. ventricosa	ov	–	?	SCP	Gutknecht (1967a)
Griffithsia pulvinata	cv	vo	ov	N	Findlay, Hope & Williams (1969)
Acetabularia mediterranea	co	co	oc	N	Saddler (1970)
Valoniopsis sp.	cv	Q	Q	N	Findlay et al. (unpublished)

[a] Q = electrochemical equilibrium
ov = medium to vacuole
oc = medium to cytoplasm
etc.

[b] N = Nernst potential, steady-state
FR = Flux-ratio
SCP = Short-circuited and perfused, steady-state
? = possible artifact (see text); no active transport?

124

sometimes coupled, more or less, with that of potassium inwards. The mechanism is possibly electrogenic;

(ii) a sodium/chloride pump; sodium ions are exchanged for ammonium ions or protons, and, coupled to this, chloride ions are exchanged for bicarbonate ions or hydroxide ions;

(iii) a chloride pump;

(iv) a proton pump (gastric mucosa);

(v) a potassium pump, which is electrogenic.

As we shall see, nearly all of these systems may find a parallel in algal cells. Let us consider some of the results for giant algal cells, which are summarised in table 9.1. The most consistent finding is that there is active transport of sodium out of the cell from cytoplasm to exterior. This depends on the knowledge, derived from flux measurements (table 6.1) that the outer membrane is not impermeable to sodium. However the small values sometimes observed for sodium fluxes suggest that the necessary wait for a steady-state could be rather long. Thus Hope & Walker (1960) found $1.3 \text{ nmol m}^{-2} \text{ s}^{-1}$ for ϕ_{ov}, and Barr (1965) found $0.8 \text{ nmol m}^{-2} \text{ s}^{-1}$. For cells whose vacuole volume is $3 \times 10^{-8} \text{ m}^3$, the equilibration half-time would be of the order of 1700 h (or 10 weeks). For marine cells with large fluxes, the problem is less acute. Only in *Valoniopsis* is it easy to envisage no active transport of sodium, with an expected internal concentration of 600 mM and an observed average of 620.

Apropos of the steady state approach, an interesting instance is offered by Ca^{2+} in *Nitella* (Spanswick & Williams, 1965). The vacuolar concentration is of the order of 10 mM, the expected (equilibrium) concentration, 13 M; yet the influx is so small ($0.5 \text{ nmol m}^{-2} \text{ s}^{-1}$) that no outward pump need be postulated for cells which have grown to their observed size in less than 10 weeks. The equilibration half-time, in the absence of active transport, is some 200 years!

In many cells (table 9.1) there is a consistent indication of inward potassium transport. Various other fluxes have been postulated to be coupled with the active potassium influx and sodium efflux, such as a light-stimulated chloride influx, or a proton efflux. It seems fair to say that ideas of these couplings may be expected to change as the mechanisms are elucidated.

In all the freshwater plants here considered, and in most of the marine ones, an inward transport of chloride must be postulated. Commonly the net influx of chloride occurs only or chiefly in the light, and much work has been devoted to the nature of this connection, still perhaps not clear. The remainder of this chapter is devoted to discussing the active transport of sodium and potassium.

Na^+/K^+ linked transport

The best-known active transport system in animal cell membranes is probably the coupled K^+ influx and Na^+ efflux found in nerve axons, red blood cells, muscle cells and many others. The system is characterised by:

(i) dependence of Na^+ efflux on $[K^+]_o$,

(ii) inhibition by the cardiac glycoside ouabain and related substances, when applied to the outside of the membrane.

(iii) a requirement for ATP on the inside of the membrane,

(iv) a ratio between the rates of Na^+ coming out and K^+ going in, between $3:2$ and $1:1$, and

(v) a stoichiometry between ATP hydrolysed and Na^+ actively transported of about $1:3$.

When an ATPase was found that was stimulated by K^+, Na^+ and Mg^{2+}, and was also ouabain-sensitive (see Skou, 1965), the world had acquired its first putative ion transport enzyme after years of anticipation that membranes should in fact harbour such enzymes. The incidence of the ATPase in animal tissues corresponds very well to the occurrence of K^+/Na^+ linked transport in those tissues. The dependence of the rate of ATP hydrolysis by the ATPase on the concentrations of K^+, Na^+, Mg^{2+} and ouabain, and many other correlations, leave little doubt that the enzyme is closely associated with K^+/Na^+ transport. Bonting(1970) gives a thorough review of these properties.

It was natural, after this discovery, to search for similar phenomena in plant cells with a presumptive active efflux pump for Na^+, and active influx pump for K^+. We shall see how far this has been successful.

The best case to be made out for a linked K^+/Na^+ pump, similar to that found in animal cells, is for *H. africanum* (Raven, 1967, 1968*b*). In these cells:

(i) the light-stimulated fraction of Na^+ efflux is entirely $[K^+]_o$-dependent, increasing from zero to a maximum 'saturation' level of 6 nmol $m^{-2} s^{-1}$ as $[K^+]_o$ increases from 0 to 0.5 mM;

(ii) the ouabain-sensitive part of Na^+ efflux is related to, but usually less than the ouabain-sensitive fraction of the K^+ influx in the light. This ratio is about $2:3$. Concentrations of 0.5–1 mM ouabain need to be used. Some cultures of *H. africanum* are not ouabain-sensitive (Raven, 1970c), but those that are, exhibit the behaviour being described;

(iii) the ouabain-sensitive part of the K^+ influx and of the Na^+ efflux disappear in the dark or at low temperatures.

The above is probably sufficient evidence to support the case, even though it is necessary to use much higher concentrations of ouabain than those that inhibit ion transport in animal cells, namely 10^{-5} — 10^{-4} M. We may note that one of the few other instances of ouabain-inhibited ion transport in plant cells is in guard cells of stomates, where closure depends on inward potassium transport, and 10^{-5} M ouabain causes 50% slowing (Thomas, 1971).

Amongst other giant algal cells, potassium influx in *N. translucens* is often but not always ouabain-sensitive. In some early experiments, the K^+ influx of these cells was reported to drop from 8–9 nmol $m^{-2} s^{-1}$ to about 3, in the presence of 5×10^{-5} M of the glycoside (MacRobbie, 1962). No effect on Na^+ efflux was mentioned. *Tolypella intricata* and *C. corallina* were unaffected by up to 1 mM ouabain, neither K^+ influx nor Na^+ efflux being altered (Smith, 1968a; Findlay *et al.*, 1969). The efflux of Na^+ was also unaltered by reducing $[K^+]_o$ to zero. Instances of a genuine ouabain inhibition of membrane fluxes in plant cells seem to be rare: though there is no reason to think that ouabain would certainly find its way to the site of active potassium transport if this were at the tonoplast. This is a possible explanation of the lack of effect of ouabain in *Chaetomorpha*, where active K^+ transport may be more important at the tonoplast. However, an ATPase with the correct properties of being stimulated by K^+, Na^+ and Mg^{2+} has not been located in plant cells despite a diligent search (e.g. Atkinson & Polya, 1967).

Cation transport linked to that of chloride

Part of the light-stimulated potassium influx in *Hydrodictyon* is insensitive to ouabain and is associated with light-stimulated chloride influx but not with sodium efflux (Raven, 1968b).

Thus, in the light, chloride influx is increased by the presence of K^+ in the medium, and, reciprocally, an increased K^+ influx is produced by Cl^- in the medium. Na^+ will also promote Cl^- transport in the light, the concentration required for half maximum stimulation being about 0.5 mM as compared with 0.1 mM of potassium.

There is a residual K^+ influx in dark conditions, or when zero-chloride solutions containing ouabain are used. This residual influx is generally interpreted as a passive influx, its magnitude being $1-4$ nmol $m^{-2} s^{-1}$.

A link between K^+ and Cl^- influxes is also found in some batches of *Chara corallina*. It is impossible to make a general statement about light-stimulation of fluxes in *C. corallina* in the face of the unfortunate variability of this stimulation; in this as in many other cells, respiration, photosynthesis and glycolysis may have varying importance in supporting ion transport under conditions that have yet to be clarified. No doubt the old practice of measuring fluxes in unstirred, un-buffered solutions contributed to this variability.

The linkage of K^+ and Cl^- influxes is not, in the genera examined, in a $1:1$ ratio. Hence, a strictly neutral pump, that causes equal numbers of $K^+ + Na^+$ and of Cl^- to be transported into the cytoplasm per unit time, is not a sufficient explanation. In any case, the light-dependent parts of K^+ and Cl^- influxes may be unequally inhibited or stimulated. Further, imidazole and other bases inhibit K^+ influx and stimulate Cl^- influx (chapter 10). Of course, an incorrect impression of the flux balance may have been obtained if influxes have been under-estimated in some instances (see chapter 6), or fluxes omitted from the scheme altogether (H^+).

Energy sources for K^+/Na^+ transport

We saw above that there is a similarity, of the light-stimulated part of potassium influx not linked to chloride influx, to K^+/Na^+ transport in some animal cells. In view of this it is natural to enquire whether ATP is also the energy source in *Hydrodictyon* and *N. translucens*.

Experiments designed to answer this question are of several types. Some involve the use of uncouplers and inhibitors of electron transfer in photosynthesis and of photophosphorylation. The intention is to find what partial reactions in photosynthesis might be closely associated with K^+ transport. The same approach has been taken with the problem of the light-stimulated influx of anions (chapter 10). Other experiments involve the interaction of light intensity, wavelength and carbon dioxide concentration with K^+ influx, with the same object of examining correlations between partial reactions in photosynthesis and K^+ transport. In order to discuss these experiments, it will be necessary to refer to details of the mechanisms of photosynthesis and the sites of action of the chemicals used, in so far as they are known. The 'Z-scheme' (fig. 9.1) for photosynthetic reactions has stood up well as a

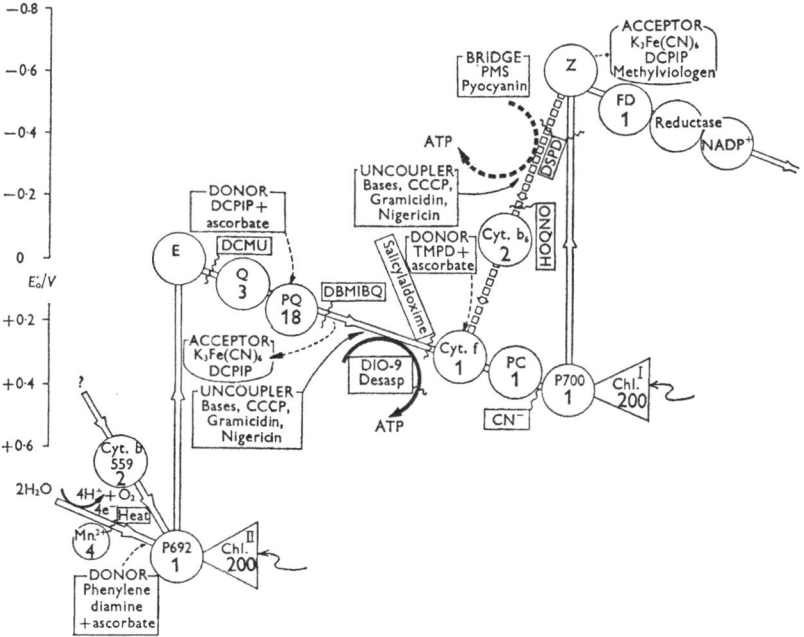

Fig. 9.1 Postulated components of the photosynthetic apparatus; pathways of electron flow; places of action of artificial electron donors and acceptors, of uncouplers, electron transport inhibitors and phosphorylation inhibitors, arranged on a scale of standard redox potentials. The enclosed numerals refer to the numbers of molecules of the substance per photosynthetic unit. Z and E are unknown primary electron acceptors, FD = ferredoxin, Cyt = cytochrome, PC = plastocyanine, PQ = plastiquinone, Q = quinone, TMPD = tetramethylphenylenediamine, DBMIBQ = Dibromomethylisopropylbenzoquinone; P692 and P700 are molecules of chlorophyll constituting reaction centres in the photosystems; other abbreviations are explained at the beginning of the book.

working hypothesis, and we may use it without being committed to its permanence.

Our understanding of K$^+$ transport in *H. africanum* rests heavily on the extensive work of J. A. Raven (1967, 1968*b*, 1969*a* & *b*, 1971). He found the potassium influx in *Hydrodictyon* to have the following properties, all of which are attributed to the Na$^+$-linked, ouabain-sensitive fraction of the influx, with experimental justification.

(i) It is stimulated by white light. 'Far' red light of wavelengths between 700 and 740 nm is more effective, relative to red light at 680 nm, in stimulating K$^+$ influx than it is in stimulating carbon dioxide fixation. The light stimulation is not inhibited by concentrations of DCMU and HOQNO which do inhibit carbon dioxide fixation. Higher concentrations of both substances do inhibit both processes: the differential effect is less clear for HOQNO than for DCMU. These substances are accepted as inhibitors of non-cyclic electron flow (fig. 9.1). The results are consistent with a dependence of K$^+$ influx on ATP; and with cyclic electron flow as a producer of such ATP (but not of NADPH$_2$ for carbon dioxide fixation) in the presence of far red light only, or of white light with just sufficient concentrations of DCMU or HOQNO.

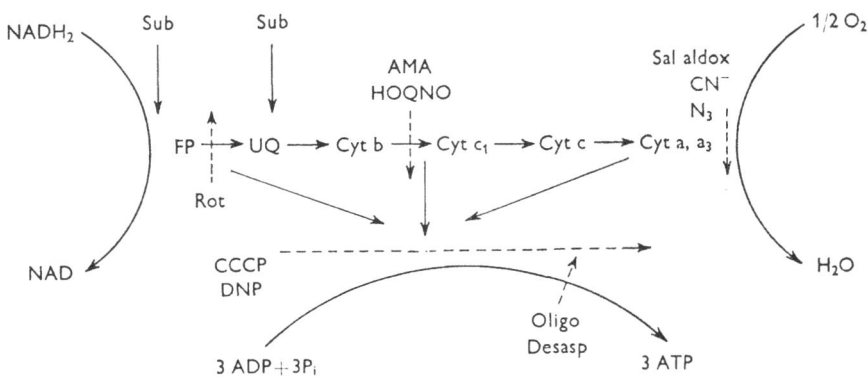

Fig. 9.2 The currently-accepted scheme for electron transport (e–t) during respiration, from NADH$_2$ to oxygen. Also shown are sites of entry of substrate (Sub) and of action of inhibitors of e–t, e.g. rotenone (Rot), of phosphorylation [oligomycin (Oligo) and desaspidin (Desasp)], and of uncouplers of phosphorylation (CCCP and DNP). The 'sites of coupling' of phosphorylation are supposed to be three in number and to be located approximately as shown. AMA = antimycin A, Sal aldox = Salicylaldoxime.

(ii) Its light-stimulation is less sensitive to CN^- than is carbon dioxide fixation, and less sensitive than the dark influx. This is consistent with CN^- acting as shown in figs. 9.1 and 9.2, to inhibit oxidative production of ATP and carbon dioxide reduction, but not photophosphorylation. These hypotheses are then consistent with the K^+ influx depending on photo ATP in the light and oxidative ATP in the dark.

(iii) Its light-stimulation is inhibited about equally with carbon dioxide fixation by photophosphorylation uncouplers (CCCP, DNP, desaspidin), by electron transport inhibitors which inhibit cyclic as well as non-cyclic flow (DSPD, salicylaldoxime, antimycin A), by energy transfer inhibitors (DCCD, arsenate) and by an ATP-trapping substance (ethionine). These results are consistent with the light-stimulation being produced by ATP.

(iv) Its light-stimulation can be shown to be less in the presence of carbon dioxide under the following conditions: no external chloride, no oxygen present, low light intensity. The carbon dioxide concentration affects the light stimulation in the same concentration range as it affects carbon dioxide fixation: presumably carbon dioxide is here competing with K^+ influx for a common energy supply.

(v) Its light-stimulation is much more sensitive to Dio-9 and DCCD than is carbon dioxide fixation, suggesting specific inhibitions of the K^+ transport mechanism.

(vi) Its value in the dark may still be reduced by a number of substances which uncouple or inhibit the oxidative production of ATP (CCCP, HOQNO, antimycin A, salicylaldoxime) whose sites of action can be identified on fig. 9.2. CN^- and DNP inhibit markedly, the former at much lower concentrations than are required to inhibit the light stimulation. Of these substances five have been shown to inhibit (CN^-, HOQNO, antimycin A) or to uncouple (DNP and CCCP) respiration in *Hydrodictyon*. These results are consistent with the dark influx being powered by oxidatively produced ATP.

This extraordinary body of results is persuasive evidence of the role of ATP in driving the ouabain-sensitive K^+ influx. Although most of the inhibitors used affect both photo- and oxidative phosphorylation, the conclusion is strongly indicated that in the light, the required ATP is produced by photophosphorylation, which may be cyclic or non-cyclic, while in the dark it is produced by oxidative phosphorylation. There are many good things about Raven's work, not least his recent study of the respiratory effects of most of the inhibitors he has used which demonstrates the penetration of these substances into the cells. But problems remain. The argument that K^+ influx may be supported by cyclic photo-phosphorylation in light of 710 nm is not for example supported by independent evidence of the existence of such cyclic flow *in vivo* (either in far red light or when carbon dioxide is absent): the use of K^+ influx as an assay for cyclic flow thus represents an argument about which many will wish to suspend judgement. Much of the discussion hangs on sites of action of inhibitors which tend to be less than perfectly known, and to have been studied with organelle preparations *in vitro*. Their effects in the whole cell have only been studied in terms of carbon dioxide fixation, K^+ and Cl^- influx, and respiration in the dark. A clearer picture might emerge if oxygen evolution measurements were available; one does need to know whether photophosphorylation is being uncoupled or inhibited, and many of the substances used have both actions *in vitro*, though at different concentrations.

Given the complex of control systems in the intact cell, it may not be adequate to make simple arguments based on systems without feedback. Thus the few experimental measurements of ATP:ADP ratios in plant cells have shown no dependence on light or inhibitors except ethionine, suggesting effective control. Heber (personal communication) has found threefold changes in ATP level in leaves between dark and light, using a very rapid freezing technique. Such changes may occur in *Hydrodictyon*, or one might invoke a rather mystic compartmentation in *Hydrodictyon* that makes photo-ATP much more effective in K^+ transport than oxidative ATP. Given that mitochondria are supposed to produce ATP for export, and chloroplasts for home consumption, this preference for photo-ATP would be surprising. The low intensity of light at which K^+ influx saturates in 1 mM CO_2 (Raven 1969*b*) is in fact close to the compensation point, where the rate of production of ATP from oxidative and photo sources is less than twice the dark rate: yet the K^+ influx is at five times the dark influx. Whether polyphosphate acts as an ATP:ADP buffer in plant cells is a further source of uncertainty in the interpretation of changes in active fluxes. Work on the transport of ATP across the intact chloroplast envelope is relevant here, especially if the transport is coupled to that of $NADH_2$; but the results do not justify an extended discussion.

Apparently unlinked K^+ transport

Blount & Levedahl (1960) seem to have been the first to apply the 'short-circuit and perfusion' technique (Ussing & Zerahn, 1951) to discover the active fluxes in a plant cell. In *Halicystis ovalis*, the vacuole may be perfused with sea water, the open-circuit p.d. then being about -80 mV. The whole of this appeared across the plasmalemma, which makes *Halicystis* a little unusual amongst the marine coenocytes (see table 5.4). When the protoplasm as a whole is short-circuited, the p.d. across neither membrane is, generally speaking, zero. However, this does not affect the thermodynamic argument that those ions which carry the short-circuit current (SCC) must be actively transported since the electrochemical potential difference between the medium and vacuole has been reduced to zero by perfusing and short-circuiting. However, cytoplasmic concentrations must come into a steady-state at the new membrane potentials (produced by short-circuiting) before steady currents can be attributed to active transport. The conclusion from the experiments with *Halicystis* was reassuringly simple – there was a net efflux of Na^+ and net influx of Cl^-, which together accounted for the SCC. However, the influxes and effluxes of Na^+ and Cl^- were measured in four different experiments where, unfortunately, the SCC ranged from $0.1–0.5$ A m^{-2}. The conclusions regarding the fluxes and the SCC were arrived at by comparing the percentage of the appropriate SCC carried by the four different unidirectional fluxes. Potassium ions are apparently at electrochemical equilibrium in the normal cell and fluxes of K^+ were not measured in the perfused, short-circuited preparation.

Gutknecht (1967a) found that cells of *Valonia ventricosa* would tolerate perfusion with an artificial cell sap, with the same used as external medium, but that the cytoplasm formed itself into aplanospores if the vacuole was perfused with sea water. The artificial sap contained, amongst other ions, 618 mM K^+. Cells when perfused with this sap displayed an open-circuit potential of about $+74$ mV. The high concentration of

K$^+$ outside the plasmalemma in these experiments depolarised this membrane so that the cytoplasm was $+18$ mV ($\psi_{co} = +18$); ψ_{vc} was $+56$ mV. In the short-circuited condition, it may be estimated that $\psi_{co} = +14$ mV, and $\psi_{vc} = -14$ mV if the relative conductances are $g_p:g_t = 20:1$ (cf. *Chaetomorpha*).

The mean SCC was 94 mA m^{-2} and this value, together with the open-circuit p.d. of 74 mV, indicates an overall conductance of 1.3 S m^{-2}, in reasonable agreement with earlier measurements on the same species by Blinks (1930*a*). In the short-circuited, perfused condition, there were net influxes of each of potassium, sodium and chloride *but these did not agree with the SCC*, one half of the latter being unaccounted for.

An unusual experimental error may have influenced these results (Gutknecht, personal communication): the radioactivity of the naturally-occurring isotope ^{40}K in 1 ml of 618 mM K$^+$ may be estimated at 4×10^3 dpm. Hence, radioactivity not originating from an influx would be found in the perfusate. The error would be most serious in the determination of chloride influx since the sample counting rates were lower.

The other very significant feature of the *Valonia* results of Gutknecht is that the open-circuit p.d. and the SCC were much less in dark conditions than in light of 4 klux. The open-circuit p.d. in the light is possibly an electrogenic one, since all the Nernst potentials are less than about 20 mV for the tonoplast. Even the dark value of ψ_{vc} was more than 60 mV. The firm conclusion from these experiments is that there is inward active transport of K$^+$, in agreement with earlier studies with intact cells. Later experiments (Gutknecht, 1968*a*) seem to show that as little as 10^5 Pa (~ 1 atm) pressure in the vacuole during perfusion with artificial sap (sea water outside) has the effect of reversibly reducing the influx of potassium considerably. Once again, if the samples of perfused sap taken for radioactivity measurements were less frequent during the period of pressure application, the presence of ^{40}K in the sap may have mimicked a component of the influx, the size of which would appear to be inversely proportional to the time interval over which the flux was considered. The possible effect of pressure on active transport is well worth following up, since it might constitute a feedback control factor in cell expansion. This would be a rational approach for a marine cell to adopt:

the alternative of controlling the internal concentration would need rather precise sensing of the total internal osmolarity.

In *Chaetomorpha* and *Valoniopsis*, the active transport of potassium is supposed to occur at the tonoplast and possibly at the plasmalemma as well; the potassium influx at the plasmalemma is light-stimulated, but the chloride influx is not. CCCP (at 4–7 μM) lowered both potassium and chloride influx at the plasmalemma in *Chaetomorpha* (Findlay *et al.*, 1971) but (at 6 μM) it inhibited only the flux of potassium from cytoplasm to vacuole in *Valoniopsis*.

Some of the plasmalemma influx of potassium is taken to be passive in cells of both these plants, and most of the chloride flux at both membranes can be thought of as passive. Thus the effect of CCCP may be directly upon membrane permeability (Findlay *et al.*, 1971). For this and other reasons, inhibition of a flux by this uncoupler cannot be taken as conclusive evidence for linkage to phosphorylation.

The giant marine algae seem to be ideal systems for investigating sodium and potassium transport, but are only slowly yielding their secrets. Difficulties in growing and collecting tropical genera no doubt contribute to the slow progress.

CHAPTER 10

Studies of active transport–anions

Chloride: rates of influx in light and dark

The observation that there is a light requirement for continued uptake of chloride or bromide ions into the vacuole of *Nitella* dates from the 1920s. A persistent, net influx, occurring as it does against an electrochemical potential gradient (table 5.4), requires a constant supply of energy. This is thought of as coming more or less directly from photosynthesis in many algae.

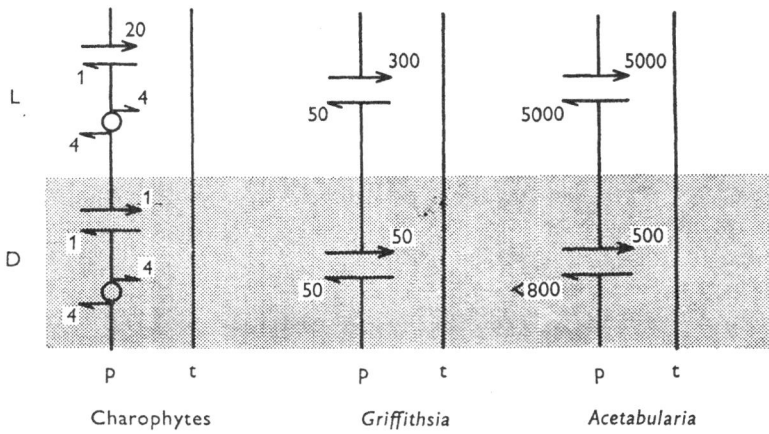

Charophytes   *Griffithsia*   *Acetabularia*

Fig. 10.1   Influxes and effluxes (nmol m^{-2} s^{-1}) of chloride, in light (L) or dark (D) conditions, averages for the plants named. A doubly-barbed arrow indicates active transport, singly, exchange-diffusion or passive efflux.

Figure 10.1 shows the observed chloride influx and efflux in light and dark, for three genera of giant cells. Plant cells may be expanding in volume or not – isolated charophyte cells rarely grow – and may be exposed to continuous light or day–night cycles. The cells will be expected to be in a steady-state of zero net flux only after growth has ceased and after a prolonged period in steady light. In such light, the respective effluxes are less than the influxes in most cells studied, but not in

Acetabularia, where there is a steady-state after several hours in the light.

Charophyte cells have been known to burst in strong, continuous light in the laboratory, but generally the control systems in plant cells are adequate to prevent disaster. In *N. translucens* elongated and very short (younger) cells have approximately the same $[Cl^-]_v$, if we can assume $[Cl^-]_v = [K^+]_v + [Na^+]_v$ (MacRobbie, 1962).

In the dark, influxes generally fall and effluxes remain as in the light, or increase (*C. corallina*: Hope *et al.*, 1966). In many cells the residual influx in the dark is too great to be explained by passive, independent diffusion. A flux-ratio ϕ_{oc}/ϕ_{co} of 10^{-3} to 10^{-4} is expected from the electrochemical gradient on the basis of equation 9.2, and hence the passive, independent chloride influx should be less than 10^{-11} mol m^{-2} s^{-1}. The observed influx may be due to an exchange-diffusion system in *Chara* (Findlay *et al.*, 1969) or it may be an active influx energised by respiration in *Hydrodictyon* (Raven, 1969*b*) and *Griffithsia* (Lilley & Hope, 1971*b*). A glycolytic energy source is also possible: such a source presumably powers cytoplasmic streaming in certain cells, as discussed in chapter 12. The efflux of chloride, by contrast, is affected by the membrane potential and by the external anion concentration in a way that suggests passive diffusion and exchange-diffusion components (Hope *et al.*, 1966).

Chloride influx and photosynthesis

Raven has shown for *Hydrodictyon* a clear connection between the stimulation of chloride influx and the absorption of light by photosynthetic pigments rather than by phytochrome or some other pigment system. The action spectrum for chloride influx (fig. 10.2) is similar to that for carbon dioxide fixation in these cells, showing the red drop for wavelengths beyond 680 nm. Presumably the action spectrum for oxygen evolution would follow that for carbon dioxide fixation, as it is known to do in other green algae.

In other genera, photosynthetic light absorption is implicated by the absence of the phytochrome red/far red effect, or more usually by the action of known photosynthetic inhibitors. In *Griffithsia*, Lilley & Hope (1971*b*) showed at increasing concentrations of DCMU a parallel, increasing inhibition of oxygen evolution and light-stimulation of chloride influx. In all other giant cell genera studied, DCMU is effective in inhibiting the light-stimulation, as are other characteristic photosynthetic inhibitors.

Thus there are parallel inhibitions of oxygen evolution, carbon dioxide fixation and chloride influx, together with the presence of the 'red drop', that is the decreased efficiency at wavelengths greater than 680 nm. These results forbid any suggestion that cyclic photophosphorylation (powered by PSI) may be the energy source (fig. 9.1). PSI alone is ruled out, in contrast to the situation with potassium influx. However, it is to go far beyond the evidence to say (as some have done) that it shows a close linkage between PSII and chloride influx.

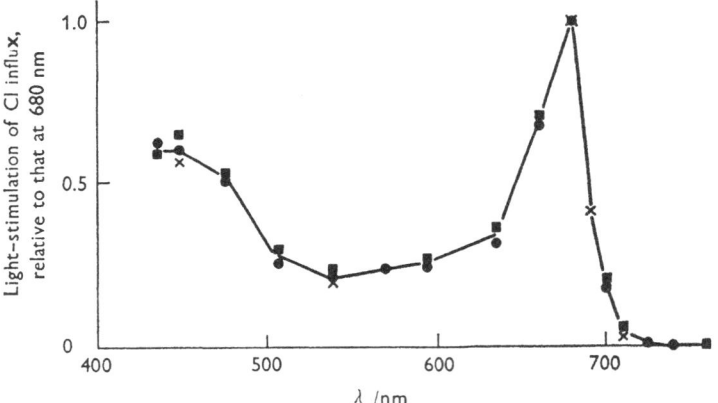

Fig. 10.2 The action spectrum for light-stimulated chloride influx in cells of *H. africanum*. The influx (light, 680 nm, minus dark) is plotted as a percentage of that at 680 nm. The symbols represent different batches of cells for which the influx at light saturation ranged from 40–48 nmol $m^{-2} s^{-1}$; dark influxes were 1–7 nmol $m^{-2} s^{-1}$. The energy flux was 0.5 W m^{-2} (from Raven, 1969a).

It is also possible to eliminate as energy source some product of carbon fixation, since chloride transport is mildly inhibited by the absence of carbon dioxide (MacRobbie, 1966; Findlay *et al.*, 1969) or not inhibited at all (Lilley & Hope, 1971b).

There are difficulties from this point on, of two kinds. First, there is some difficulty, given our ignorance of photosynthesis and of chloroplast membranes, in knowing what to postulate as a coupling mechanism between electron transport in the photosystems, and active chloride influx at the plasmalemma. The idea of a malate shuttle, or a 3-PGA–triosephosphate shuttle between chloroplasts and the cytoplasm, that results in export of energy stored in ATP and of reducing power in $NADH_2$ (Heber & Santarius, 1970; Heber & Krause, 1971) has been thought relevant to this difficulty (see Lüttge, 1973). Second, as we shall describe fully below, there is a direct conflict between the evidence from different genera, about the effects of uncouplers and inhibitors.

Chloride influx and net electron transport

The first difficulty has led to hypotheses that appear, to some at least, to be incompletely stated. Thus MacRobbie (1966) suggested that chloride influx in *Nitella* was driven by net electron transport without specifying how, except that the coupling was not via ATP. A well-known earlier hypothesis had sought to find the driving force for ion transport in charge separation reactions:

$$H \rightarrow H^+ + e^-$$
$$H_2O \rightarrow H^+ + OH^-$$

where in each reaction the charged products are thought of as appearing on opposite sides of a membrane (see Robertson, 1968). Such reactions may well occur in both photosynthesis and respiration, but these occur at membranes different from the plasmalemma, and separated from it by yet more membranes. This difficulty was never solved, but the hypothesis, an ancestor of Mitchell's theory, had the advantage of predicting stoichiometric ratios for $Cl^-:e^-$ or $Cl^-:O_2$ – the simplest values would be 1 and 4 respectively.

A direct, one-for-one connection in the giant algae between electron flow in photosynthesis and chloride influx is negated by experimental evidence. In fact the ratio between these two quantities does not seem to have a particular fixed value. This is a necessary consequence of the dependence of chloride transport, but not of photosynthetic rates, on $[Cl^-]_0$ and of their different dependences on light intensity. At intensities below 1 W m^{-2} when both processes are light-limited, Raven's results (1969*a*) give $Cl^-:CO_2$ as about 0.7; but chloride transport saturates at about 1 W m^{-2} while carbon dioxide fixation saturates at about 10 W m^{-2}: the stoichiometric ratio will vary with light intensity from 0.7 to 0.07. In *Chara* and *Griffithsia*, where rates of oxygen evolution have been measured, the results for saturating light intensity and saturating chloride concentration are:

	ϕ_{Cl}/nmol m^{-2} s^{-1}	ϕ_{O_2}/nmol m^{-2} s^{-1}	$\phi_{Cl} : \phi_{O_2}$
Chara	15–40	70–200	0.08–0.6
Griffithsia	200–700	500–700	0.3 –1.4

(Here the range of Cl$^-$: O$_2$ is obtained by dividing the extreme values, and is not the observed range)

These ratios are all much less than 4, and do not encourage the adoption of redox-pump models. Also, though oxygen evolution follows the beginning of illumination with a delay of only a fraction of a second (Joliot, 1967), chloride influx begins slowly and declines slowly when the light is switched off. In *G. flabelliformis* the time to half the maximum steady rate is 5–15 min (Lilley & Hope, 1971*b*): in *Chara* it is sometimes much longer.

The second difficulty in the chloride-transport story is the contradiction in the experimental evidence about the necessity for ongoing phosphorylation.

Photophosphorylation and chloride transport

Non-cyclic photophosphorylation is generally thought to be tightly coupled *in vivo* to net electron flow, although the exact stoichiometry is still uncertain (Walker & Crofts, 1970). Thus it might couple electron transport to chloride transport, and uncoupler experiments might decide this. These experiments have been performed by MacRobbie (1966) and Raven (1969*b*) who discount any necessity for phosphorylation to proceed at the same time as chloride transport. Fig. 10.3 shows some of the evidence for *H. africanum*; in these experiments carbon dioxide fixation and potassium influx were taken as a measure of uncoupling and of ATP level; both were much more sensitive to CCCP than was chloride influx.

The contrary hypothesis is supported by other experiments, on *Chara* (Smith & West, 1969) and *Griffithsia* (Lilley & Hope, 1971*b*), which have shown parallel effects on carbon dioxide fixation and the light-stimulated part of chloride influx. In the second genus, it was observed that with 10 μM CCCP present, oxygen was evolved at 70% of maximal rate while carbon dioxide fixation was negligible, and chloride transport, only 17% maximal. This leads to the opposite conclusion from that given above, namely, that non-cyclic phosphorylation *is* necessary for continued chloride transport.

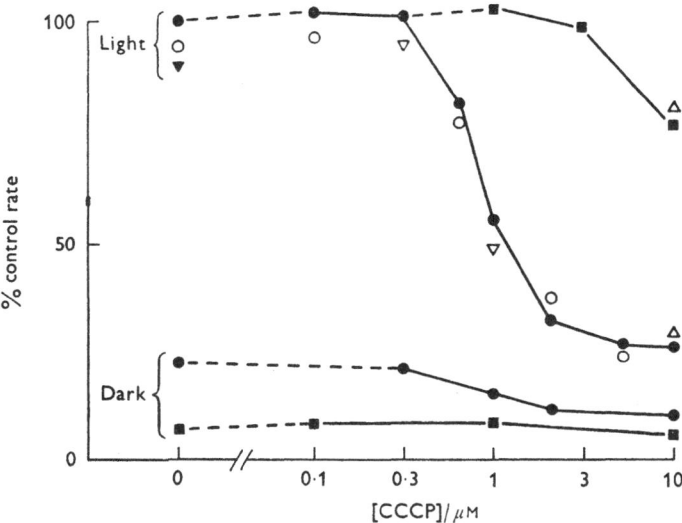

Fig. 10.3   Effect of CCCP on K^+ influx and Cl^- influx in light and dark in *H. africanum*, plotted as percentage of control rates which were: K^+ influx (air, light = ●) 8.9 ± 0.9 nmol m^{-2} s^{-1}; Cl^- influx (air, light = ■) 20.4 ± 2.1 nmol m^{-2} s^{-1}. The other symbols represent: ○, K^+ influx (N_2, light); ▽, K^+ influx (air, light, DCMU 10^{-7} M); the upright triangles are unexplained (from Raven, 1969*b*).

This disagreement might be resolved by experiments in which ATP levels are measured under a variety of experimental treatments, together with chloride fluxes. One such study (Lilley & Hope, 1971*a*) has shown ATP levels almost independent of light, dark and the presence of inhibitors and uncouplers. Only iodoacetate with darkness reduced ATP:ADP very much. This is perhaps a warning, but does not solve the present difficulty.

Chloride influx as a secondary active transport

Here as in the face of other contradictions, people have looked to Mitchell for new paradigms. Smith (1970) has suggested that a Cl^-/OH^- antiport is involved in chloride transport.

The power source then is the electrochemical gradient of hydroxide ions or of protons across the plasmalemma; we have, in the terms of chapter 9, an instance of secondary active transport. The primary active transport would on Smith's scheme be the outward pumping of protons; there may be K^+/H^+ and Na^+/H^+ antiports, or some involvement of potassium and sodium in the primary active transport.

MacRobbie (1966) had found ammonia and imidazole to stimulate chloride influx in the light and had concluded that they did so – as uncouplers – by stimulating the rate of photosynthetic electron transport. Suspicion that this was implausible was much stronger when Smith & West (1969) found that although high concentrations of imidazole would uncouple in isolated *Chara* chloroplasts, in the time and concentrations used in ion transport studies imidazole had no effect on carbon dioxide fixation in intact cells.

Smith (1970) showed that a number of bases (tris, imidazole, ammonium) would stimulate chloride influx in the light, while zwitterionic or anionic buffers would not. On his Cl^-/OH^- hypothesis, these bases act by penetrating the cell in their electrically neutral form; the neutral molecule acquires a proton in the cytoplasm, increasing the internal pH and the driving force on hydroxide ions. He further showed light-stimulation to be inhibited at high external pH. A better test of the hypothesis is to produce a chloride influx, in the dark, of about the light-stimulated level, by imposing a pH gradient on the plasmalemma. This he has done (1971); and further, bases stimulate the pH-driven influx in the dark as they do the light-driven influx.

On this model, if the exchange of chloride for hydroxide is electrically neutral, a net chloride influx can occur (Mitchell, 1967) if:

$$[H^+]_c/[H^+]_o > [Cl^-]_c/[Cl^-]_o$$

The values of $[H^+]_c$ and $[Cl^-]_c$ are scarcely known parameters: if our argument is accepted that $[Cl^-]_c$ in *Chara* is about 10 mM, at an external concentration of 1 mM, net chloride influx can be driven by an internal pH one unit more alkaline than the external. This requires a rather alkaline cytoplasm when pH_o is 8. The model also requires rather large changes in pH_c between light and dark, through a light dependence of the proton transport pump. The mode of coupling of energy to this pump is not specified so far. Perhaps a model will be sought in the bicarbonate-stimulated ATPase that may be implicated in proton or chloride transport in the gastric mucosa (Wiebelhaus *et al.*, 1971). On the credit side, the model does explain

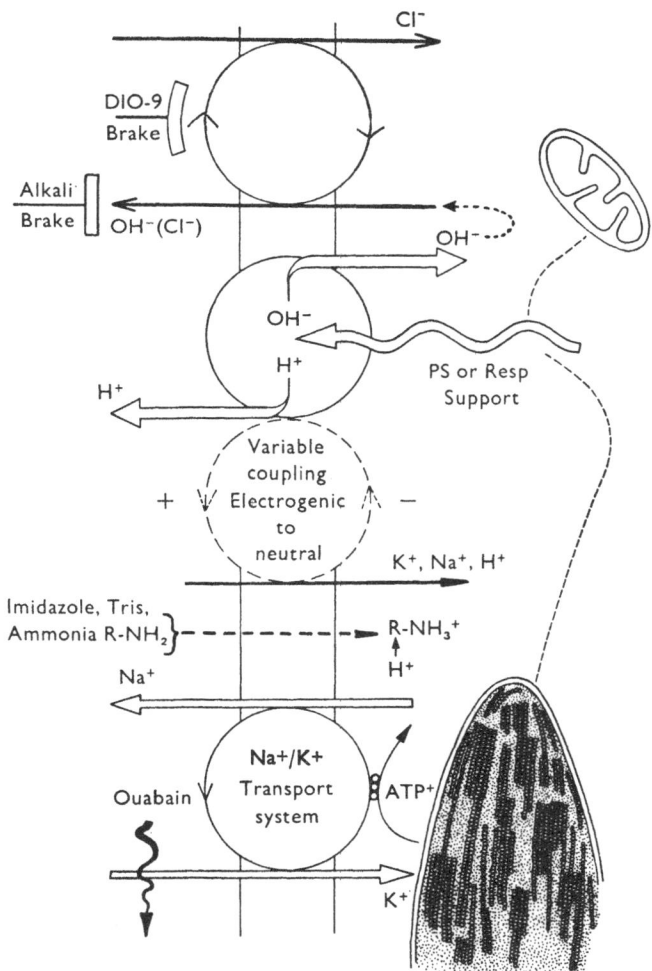

Fig. 10.4 A schematic diagram including several of the active transport schemes discussed in the text. At the bottom is the Na^+/K^+ exchange transport ATPase, postulated for *H. africanum* and *N. translucens*. Near the top is the H^+/OH^- transport mechanism that may be coupled to cation influx and OH^- efflux and Cl^- influx, as postulated by Smith (1970, 1971) for charophyte cells.

the action of imidazole in stimulating chloride influx while inhibiting potassium influx, if one can assume that the latter influx is not chemically coupled to the proton pump, but driven by a secondary K^+/H^+ antiport. Without imidazole, a constant internal pH would result from zero net flux of protons which implies equal net fluxes of mineral cations and of chloride. In the presence of a net influx of imidazole, which becomes protonated, the cytoplasm gains imidazole hydrochloride, rather than potassium and sodium chlorides. Much of the foregoing discussion is illustrated in fig. 10.4.

Playing our usual critical role, we must point out that Smith's experiments, like many others, do not tell us about net fluxes of chloride, but only about the unidirectional influx. The efflux is substantial in these cells, and deserves to be measured under the same experimental conditions. An exchange of internal for external chloride, if it occurs, should be separated experimentally from net transport.

Other halide fluxes

The possibility of distinguishing between bromide and chloride chemically led Hoagland & Davis (1923) to their pioneer study of bromide uptake by *Nitella*. The cells accumulated bromide in their vacuoles in the light as they were later shown to do with chloride.

Little has since been published on halide fluxes other than chloride, though [82]Br and ^{131}I are sometimes used as impromptu tracers for chloride, as ^{86}Rb used to be, for potassium. The best use of this possibility has been that of MacRobbie (1971), who succeeded in getting two points on the chloride influx time course for each cell using [36]Cl and [82]Br. In the process, she showed that, from a solution containing 0.6 mm Br^- and 0.6 mm Cl^-, and in the light, the influxes of bromide and chloride were 2.8 and 8.6 nmol $m^{-2} s^{-1}$ respectively, a ratio of about 0.3. Once in the cell, however, the fraction in the vacuole is the same for both halides at both short and long times, an interesting result. It implies, in terms of our model of fig. 6.5a, that the fluxes ϕ_{pl} and ϕ_t have equal selectivity ratios for chloride and bromide:

$$\phi_{tCl} : \phi_{tBr} = \phi_{plCl} : \phi_{plBr}$$

In terms of the MacRobbie model (fig. 6.5b) the flux ϕ_c does not distinguish chloride from bromide.

There are no published data on iodide fluxes, and no Eisenman sequence (see Diamond & Wright, 1969) for any halide-involving process in giant algae has yet been determined. In the absence of chloride outside, the transport mechanism is content with bromide, and the maximum influx is equal to that observed with chloride alone. In other words, K_m is different but V_{max} is the same, for the porter/ion interaction with these two species.

Uptake and distribution of phosphate

That phosphate movement to the vacuole involves an active transport may be shown by a consideration of the fluxes, the p.d. and internal and external concentrations. Under steady-state conditions the vacuolar sap of *N. clavata* may contain 2–4 mM phosphate and the pond water only 0.8 μM (table 5.2). The expected internal concentration at equilibrium is then 0.8 nM for $H_2PO_4^-$ and 0.8 pM for HPO_4^{2-}: obviously active transport inwards is necessary.

From another point of view, using *N. translucens* it has been observed that with more than 0.2 mM $H_2PO_4^-$ in the medium, there can be a net influx into the vacuole of about 10 nmol m^{-2} s^{-1}. This would cause the vacuolar concentration to rise by 0.16 mM per hour, in cells 0.9 mm in diameter. Since the equilibrium concentration for passive permeation is only 2 μM active transport is again inferred (see Smith, 1966). If the cytoplasmic concentration of free phosphate ion were to be kept very low by its prompt incorporation into organic phosphates, it would be just conceivable that the influx of phosphate at the plasmalemma is passive. Between cytoplasm and vacuole, however, there would exist an uphill electrochemical gradient and the flux into the vacuole would have to be active.

Figure 10.5 shows the uptake of phosphate into *N. translucens*. The total is seen to rise linearly with time; cytoplasmic activity reaches a quasi-steady state after about six hours while the vacuole acquires activity after a delay of approximately 30 minutes, and then at a rate accelerating to that into the whole cell. This pattern is typical of that expected for a series two compartment model where mixing occurs in the cytoplasm with inactive phosphate after which activity accumulates in the vacuole. However, phosphate also enters organic compounds by phosphorylation reactions, and it is not clear whether more extensive experiments would need more compartments to model them.

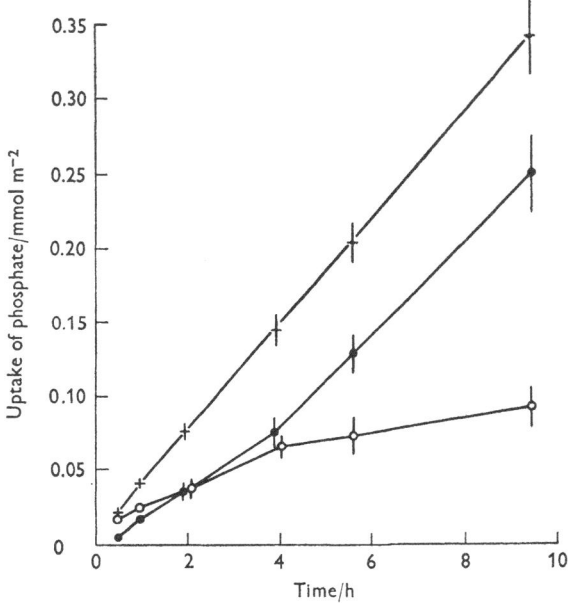

Fig. 10.5 The uptake of phosphate in the light by cells of *N. translucens*: O, cytoplasmic radioactivity; ●, that in the vacuole; +, whole cell. 5–9 cells per group, $[H_2PO_4^-] = 0.2$ mM, temperature 21 °C (from Smith, 1966).

The influx as a function of external concentration of phosphate shows the commonly observed saturation of rate, in this case beyond about 0.2 mM $H_2PO_4^-$. In freshly-collected *N. translucens* cells this maximum influx may be up to 30 nmol $m^{-2} s^{-1}$ in the light. Dark conditions usually reduce the influx to about 60% of values in the light. The effects of other conditions are summarised in the diagram fig. 10.6. Metabolic effects on phosphate influx are reminiscent of those on potassium influx (that fraction linked to sodium efflux, not to chloride influx) in *H. africanum* and *N. translucens*. Thus there is stimulation by light and operation of PSI on its own will provide this – see the results using 'Filter 2'. Uncoupler

146

concentrations which are effective in inhibiting carbon fixation reduce the light influx to much below the dark level, so it may be inferred that ATP is necessary for the active transport and may be provided either by oxidative or photophosphorylation. This finding is easier to understand than the corresponding one for potassium influx.

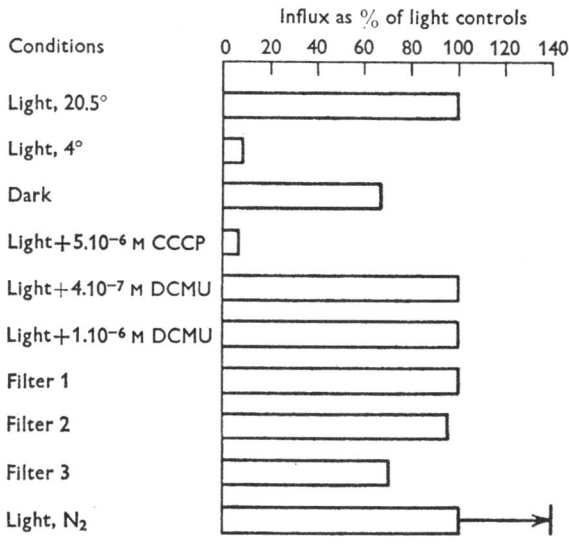

Fig. 10.6 The uptake of phosphate ($H_2PO_4^-$) under various conditions as percentages of control values (from Smith, 1966).

Using cells of *C. corallina*, the fraction of radioactivity, originating from ^{32}P in $H_2PO_4^-$, that finds its way into organic and inorganic pools has been investigated (Lilley & Hope, unpublished results). By this means it was hoped to get a measure of rates of photophosphorylation and of the effects of supposed uncouplers. The procedure would probably allow only a minimum estimate of photophosphorylation because of turnover of phosphate groups between compounds. Control influx was about 4 nmol m^{-2} s^{-1} from 0.1 mM $H_2PO_4^-$ and 60% of the activity was found in organic fractions after 60 minutes. While interpretation of such experiments is not a trivial problem, it does appear surprising that giant algal cells have not been the subject of more experiments aimed at elucidating phosphate metabolism. Such investigations are

F

overdue in view of the necessarily indirect inferences drawn from use of phosphorylation uncouplers and inhibitors.

In the dark, phosphate fluxes in *H. africanum* are inhibited to about 65% of light values. From solutions of 0.1 mM $H_2PO_4^-$ at pH 6 the influx is about 6 nmol $m^{-2} s^{-1}$. This influx is not inhibited by DCMU ($< 1 \mu M$) but is CCCP-sensitive, being inhibited to about 10% of control in 5 μM CCCP; carbon fixation is reduced to less than 10% of control rates by this concentration of CCCP. Arsenate has a prompt inhibitory effect on the influx of phosphate although it may take up to ten hours to halve the control rate of carbon fixation. This may be taken to indicate the possibility that arsenate and phosphate compete for ion transport sites on the outside of the plasmalemma.

The conclusion above, common to several genera, that phosphate transport can be light-stimulated even when PSII is inhibited, has been reached also for *Elodea* and *Ankistrodesmus*.

Sulphate

Sulphate is often regarded as a relatively inert and impermeant ion. However, there is an appreciable influx of sulphate in three or four genera of the *Charophyta* examined. Curiously, the influx is transiently *higher* in a dark period following continuous light for several hours (Robinson 1969a, b). Arguments similar to those given for phosphate show that sulphate ions must be actively transported inwards, probably at both plasmalemma and tonoplast. Reference again to table 5.2 shows that Hoagland & Broyer found about 17 mM sulphate in *Nitella* sap many years ago. The concentration is similar in *C. corallina*.

TABLE 10.1 *Mean fluxes of sulphate and rate constants for cytoplasm exchange, in cells of C. corallina, from Robinson (1969a, b)*

	ϕ_{oc}/nmol $m^{-2} s^{-1}$	ϕ_{ov}/nmol $m^{-2} s^{-1}$	Fraction in vacuole	$10^4 k$/s^{-1}
L	0.69 ± 0.02	0.29 ± 0.05	0.15–0.61	5.3 ± 0.8
D	2.0 ± 0.2	0.4 ± 0.06	0.08–0.45	1.9 ± 0.2

The fluxes in columns 1–2 were calculated from increases in radio-activity, in whole cells, or in the vacuole, in the time interval 0.5–1 h. Rate constants were calculated from the series compartments model (chapter 6). The pretreatment was 0.5 h in inactive SO_4^{2-} APW.

Robinson's measurements of the proportion of activity in vacuole and cytoplasm have been used to calculate several parameters and table 10.1 sets out mean values of fluxes at plasmalemma and tonoplast from an experiment with 20 cells of *C. corallina*. When the distribution of sulphate was studied as a function of time, the vacuolar content of sulphate showed the lag expected of the series, two-compartment model (fig. 10.7). In these experiments, the influx was similar to that found by Robinson, as was the distribution between vacuole and cytoplasm after 30 minutes and 60 minutes. While fig. 10.7 shows results for which the two-compartment model is adequate, it is not known how rapidly sulphate enters metabolic pathways.

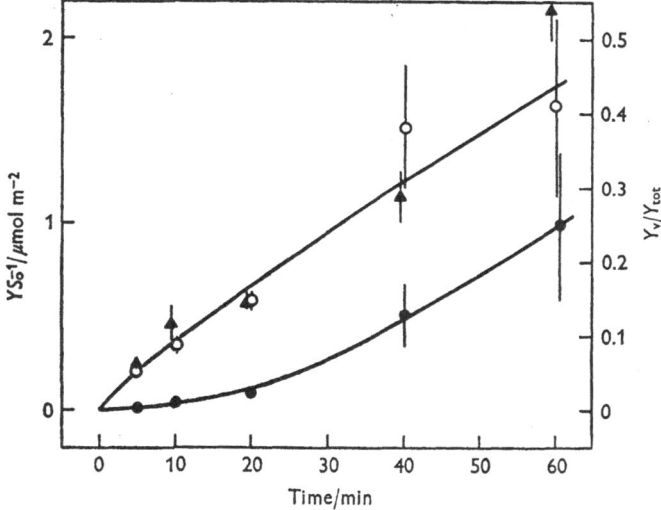

Fig. 10.7 The uptake of sulphate by cells of *C. corallina* as a function of time: ●, vacuolar activity; ○, cell total; ▲, Y_v/Y_{tot}. The points are means ±SEM for 8–10 cells, temperature 22 °C, $[SO_4^{2-}] = 0.05$ mм (Findlay *et al.*, unpublished).

Robinson investigated effects of light and the usual inhibitors on sulphate influx. Many of the cells were mature, stored internodes with low chloride influx and low carbon fixation rates. In such cells, increasing intensity of monochromatic light of wavelength 475, 539 or 660 nm caused a decrease from the transient, high, dark influx to the steady, lower light value. With 0.6 μM DCMU, carbon fixation was very small, and sulphate influx had reached the dark value. CCCP influenced only the influx in the dark, inhibiting at 5 μM. Carbon fixation in the light was inhibited by 5 μM CCCP, but the light influx of sulphate was virtually unaffected by 10 μM CCCP. 5 μM DNP caused 50% lowering of the dark influx but only much higher concentrations decreased the light influx. Cyanide at 0.1 mM caused 50% inhibition of influx in both light and dark.

From these observations it is clear that the sulphate influx behaves quite differently from the influxes of chloride and phosphate. Photophosphorylation is not the energy source in the light because CCCP is without effect. However, ATP appears necessary during dark conditions since uncouplers then reduce the flux. As with the other ion transport systems, there is the possibility of direct effects on the mechanism as well as on the supply of energy. In other algae and higher plants, experience has been that sulphate uptake may be either light-stimulated, as in *Hydrodictyon, Chlorella, Scenedesmus,* and *Elodea,* or dark-stimulated as with *Vallisneria, Thuidium, Crassula,* wheat and barley. The results with sulphate suggest control systems which react to light and chemicals in more complex ways than are postulated for potassium and chloride.

CHAPTER 11

The transport and use of carbon dioxide and bicarbonate ions

Introduction

Many of the experiments seeking to find the details of photosynthetic mediation of active ion transport have involved measurements of the rate of carbon fixation. This has often been taken as a suitable indicator of photosynthetic rate, but of course many factors combine to make carbon fixation the end-product of the light and dark reactions. There seems to be no simple absolute measure of the net rate of electron transport in whole cells. However, it is relatively simple to measure the rate of carbon fixation accurately and it is useful that much data has emerged concerning the use of carbon dioxide and bicarbonate as exogenous sources of carbon, and the effects of uncouplers and brakes of energy coupling in photosynthesis.

More importantly, it has been shown that bicarbonate ions are almost certainly actively transported into the cells before use as substrate in carbon fixation. Some of the conclusions regarding this system may be applicable to other plant cells known to be able to utilise bicarbonate directly as a carbon source.

This and other work on the importance of cyclic and pseudo-cyclic photophosphorylation justifies the use of the giant algal cells for close study of photosynthesis itself; we can expect further profitable investigations along these lines.

Rates of carbon fixation

Several dissimilar genera of giant algal cells have been found to have rather similar fixation rates, expressed on the basis of surface area, when given saturating light and carbon dioxide. Table 11.1 summarises these findings. The rates are expressed on a chlorophyll basis by using figures for chlorophyll (Chl) per unit area supplied by A. W. D. Larkum. The rate of 4.10^{-3}

F*

mol C $(\text{mol Chl})^{-1} \text{s}^{-1}$ found for *Nitella* is some 10% of the highest rates recorded in leaves; thus the giant-celled algae are in the middle of the range of carbon fixation rates (see Walker & Crofts, 1970). Dark fixation rates are very small, 0.1–2 nmol $\text{m}^{-2} \text{s}^{-1}$.

TABLE 11.1   *Rates of carbon fixation by giant algal cells*

	Area basis/ nmol $\text{m}^{-2} \text{s}^{-1}$	Chlorophyll basis/s^{-1} [a]	Reference
N. translucens	320–360	4.10^{-3}	Smith (1968*b*)
C. corallina	400		Smith (1968*b*)
T. intricata	420		Smith (1968*b*)
Nitellopsis obtusa	330		Smith (1968*b*)
H. africanum	300		Raven (1968*a*)
C. corallina	280		Smith & West (1969)
G. pulvinata	200–500		Lilley & Hope (1971*b*)

Conditions: Freshwater algae: pH 5.7–6.5, 0.2–1 mM free CO_2, 21–23 °C.
    Griffithsia: ASW, $[\text{HCO}_3^-] = 2.5$ mM, pH = 8.2, 23 °C.
  [a] Mole CO_2 per mole chlorophyll and per second.

Carbon dioxide only, or bicarbonate as well?

Raven (1968*a*) and Smith (1968*b*) have studied the rate of carbon fixation while varying the proportion of carbonic acid and bicarbonate in the medium. These concentrations depend on the pH according to the usual equations:

$$\log_{10}[\text{H}_2\text{CO}_3] = \log_{10}[\text{HCO}_3^-] + 6.37 - \text{pH}$$
$$\log_{10}[\text{HCO}_3^-] = \log_{10}[\text{CO}_3^{2-}] + 10.25 - \text{pH} \qquad (11.1)$$

The fraction in each form, according to these equations is:

pH	5	6	7	8	9	10
H_2CO_3	0.96	0.70	0.19	2.3×10^{-2}	2.2×10^{-3}	1.6×10^{-4}
HCO_3^-	4.1×10^{-2}	0.30	0.81	0.97	0.94	0.64
CO_3^{2-}	–	–	4.5×10^{-4}	5.4×10^{-3}	5.3×10^{-2}	0.36

Both genera studied, *Hydrodictyon* and *Chara*, were capable of a high rate of carbon fixation in solutions containing 2–4 mM NaHCO_3 at pH 9.5–10, where the concentration of carbonic acid would have been below about 2 μM. At about 21 °C and with saturating light, the rate with bicarbonate was up to 100 nmol $\text{m}^{-2} \text{s}^{-1}$ compared with the rate with carbonic acid of up to 300–400 nmol $\text{m}^{-2} \text{s}^{-1}$.

Thus these genera join the list of plants able to take up bicarbonate for photosynthesis – by no means all plants are able to do so. References in Raven (1970b) will lead the reader into this subject and into the biochemistry of carbonic acid and bicarbonate fixation.

In these experiments alternative explanations for the high rates observed in solutions at pH 9.5–10 might be (a) that the affinity of the relevant enzyme for carbon dioxide might be very sensitive to external pH or (b) that the bicarbonate ion acts to buffer the concentration of carbonic acid and overcomes some diffusive rate-limiting step, presumably outside the cell membrane. Neither seems plausible. Such a large effect of pH on the enzyme affinity is improbable – there is no effect of pH_0 between 5.7 and 7.3 in *Hydrodictyon*, but by 9.5 there would need to be postulated a 100-fold increase.

If it is accepted that bicarbonate as such is a carbon source for photosynthesis, it must be asked why the rates at light saturation seem so much lower than those achieved with carbon dioxide. The possible postulates seem limited to (i) a rate-limiting diffusion step across a membrane, or (ii) an active, (light-driven?) rate-limiting reaction not involved in fixation of carbon dioxide itself. Raven showed that even under light-limiting conditions, the relative rates of fixation using carbon dioxide and using bicarbonate differed by a factor of 2.8. This seems to rule out the first postulate, (i) above, which predicts similar rates when the light intensity is truly limiting. Raven showed that under his light-limiting condition there was no effect of temperature (5 and 15 °C) on the rate, and no effect of bicarbonate concentration. It seems that postulate (ii), a separate light reaction involved in bicarbonate uptake or reduction, should be adopted. This light reaction is most plausibly involved in the uptake of bicarbonate across the plasmalemma.

The extra reaction to provide bicarbonate at internal sites, which is not necessary when carbon dioxide is available, needs the participation of PSII. Thus, there is found a lower quantum efficiency for the fixation of carbon from bicarbonate, and this is consistent with the use of light energy for bicarbonate uptake. Under light-limiting conditions, light of wavelength >700 nm also has a much lower efficiency with bicarbonate than with carbon dioxide; hence one cannot postulate that PSI alone provides for bicarbonate entry – if this were so one would expect equal efficiencies with carbon dioxide and bicarbonate in far red light. Fig. 11.1 shows the data for *Hydrodictyon* on which this argument is based.

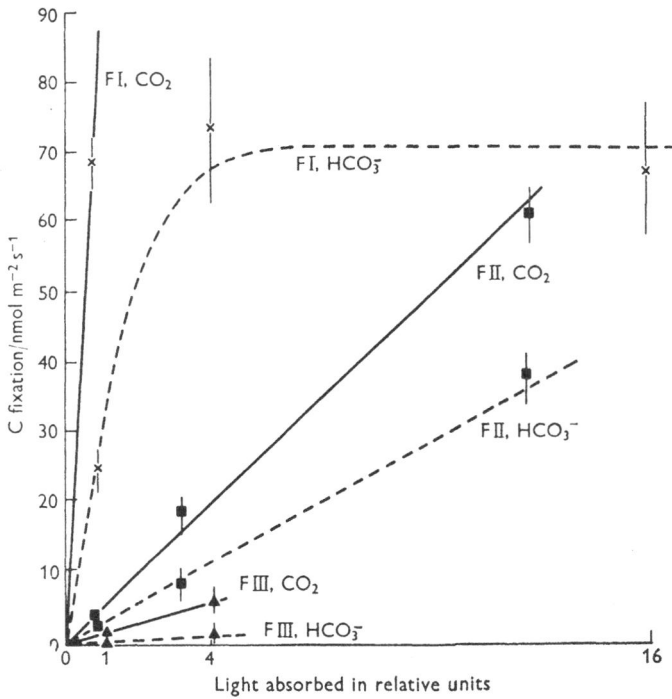

Fig. 11.1 Carbon fixation by cells of *H. africanum* under various lighting conditions, and with carbon supplied either as 1 mM total inorganic carbon at pH 7.4 (solid lines) or as 2 mM at pH 10.2 (broken lines). The fixation rate at saturating light intensity (25 units) using filter I and pH 7.4 was 300 nmol m^{-2} s^{-1}. The filters used had the following relative quantal transmissions:

	Below 730 nm	Below 705 nm	705–730 nm
Filter I	100	46	54
Filter II	40	3.1	37
Filter III	12.6	0.3	12.3

(From Raven, 1968*a*.)

The active transport of bicarbonate

If it is accepted that many of the giant algal cells are able to use bicarbonate as well as carbon dioxide, then we have to inquire about its mode of entry into the cells. There is no problem explaining the uptake of carbon dioxide if diffusion of dissolved gases through cell membranes takes place readily, as usually assumed (see Raven, 1970b). Bicarbonate may diffuse through the plasmalemma passively, or be actively transported. The evidence for its active movement has already been discussed; we can show that wholly passive entry is unlikely through a consideration of the permeability of the plasmalemma required to account for the observed rates of fixation. These rates are under-estimates of the influx of bicarbonate for reasons given below.

From the Goldman equation we can calculate the required permeability. Assuming that ψ_{co} is -125 mV, the external concentration is 2 mM and the internal concentration is negligible, we get $P = 1.5 \times 10^{-6}$ m s^{-1} for an influx of 100 nmol m^{-2} s^{-1}. The influx may be greater, or ψ_{co} more negative than -125 mV, as it is with some cells for some hours after a change to alkaline pH (the hyperpolarisation mentioned in chapter 7 and table 7.2). The passive permeability to chloride ions, which should be calculated from the efflux since the influx is active (chapter 10), is in the range 10^{-11}–10^{-10} m s^{-1}. Thus passive permeation could not result in enough bicarbonate in the vicinity of the chloroplasts to support the observed fixation rates, unless the bicarbonate permeability were many orders of magnitude greater than the chloride permeability. Once bicarbonate is in the cytoplasm it is presumed that carbonic anhydrase, present in most photo-synthesising plants, would catalyse its breakdown to carbon dioxide and hydroxide ions. Some of the carbon dioxide, produced by enzymic splitting of bicarbonate may be lost from the cell by diffusion. This suggests that the influx of bicarbonate may be rather greater than the observed fixation rate, perhaps more than twice.

The foregoing has been summarised in fig. 11.2. Appearance of hydroxide in the external medium during photosynthesis has been frequently observed, in higher water-plants for example, and Smith (1968b) showed that the medium outside C. corallina cells could rise in pH from 7 to 10 during a prolonged period of photosynthesis in a medium containing sodium bicarbonate in APW. However, the explanation of such rises in pH is not always as simple as inferred above; we can say that observation is consistent with fig. 11.2 so far.

Both bicarbonate and chloride transport are stimulated by light that produces net electron transfer through both photosystems. It has also been shown that the presence of bicarbonate ions (1 mM at pH 7.7) depresses the chloride influx by 45% in the light in *H. africanum*, so it is suggested that these ions compete for a common carrier. Chloride does not depress carbon fixation rates in bicarbonate solutions at high pH, so

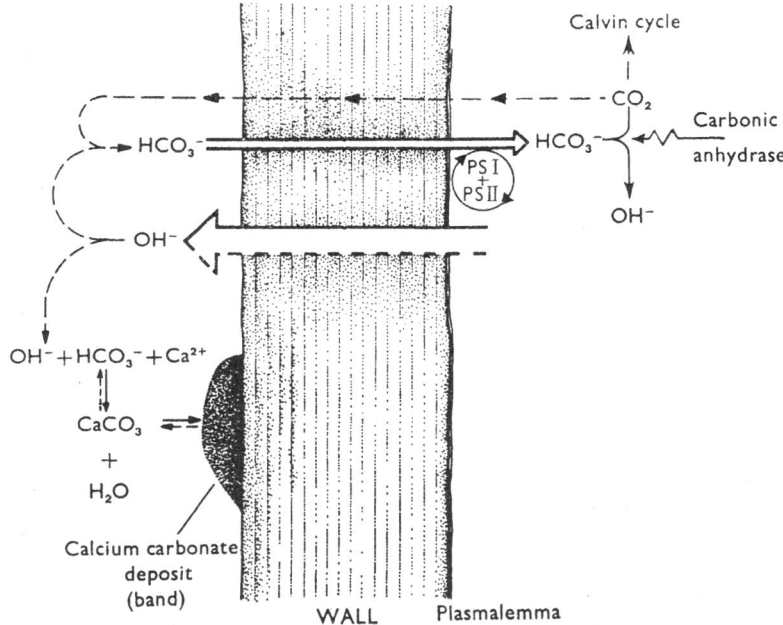

Fig. 11.2 A schematic diagram showing the active transport of bicarbonate ions across the plasmalemma, the subsequent paths of carbon dioxide and hydroxide ions and the conditions leading to formation of calcium carbonate bands, in charophyte cells.

that if there is a common carrier, its affinity for bicarbonate is much greater than that for chloride. However, the maximum rate attained by the transport system working with bicarbonate is about five times higher than when only chloride is available. This contrasts with the similar maximum rates attained by chloride and bromide. Competition between chloride and bicarbonate has been reported also in *Vallisneria* and *Chlorella* (see Raven, 1968a).

Calcium carbonate deposits and banding

It is an old observation that many charophyte plants are encrusted with crystals of calcium carbonate, which are embedded in or adhere to the cell wall. Other metal carbonates may be present, but calcium is the one commonly found. In *Nitellopsis* there is an apparently uniform deposit of scattered crystals all over the surface; in *Chara* and *Nitella* the deposit, when it occurs, is in a banded formation, the cell exhibiting lateral stripes of green and white. The deposition is due to the precipitation of calcium carbonate as the concentrations of calcium and carbonate ions reach their solubility product: the concentration of calcium is no doubt that provided by the medium, but the concentration of carbonate is clearly related to the metabolism of the cells as well as to the pH of the medium (fig. 11.2). The implication in this phenomenon is that the cell surface in charophytes is far from the uniform area assumed by physiologists. However, there were some thoughtful pioneering studies by Arens (1936, 1939), and Walker (unpublished) showed that the membrane conductance of white and green zones was identical, both in light and in dark.

Spear, Barr & Barr (1969) took the original step of placing cells with no visible carbonate deposit in a solution of phenol red: this useful indicator is apparently impermeant, and exhibits discernible colour changes over a wide range of pH (say 5.0 to 8.5). They showed (plate 7) that outside the cell, in the light, there were bands of acid and alkaline solution. The effect vanishes in the dark, or in strong buffer solutions.

Lucas & Smith (1973) showed that bicarbonate is needed in the medium for the alkaline bands to develop: they regard the whole cell surface as making the external medium weakly acidic, and the alkaline banding as a phenomenon superimposed on this. Both acid and alkali export were light-requiring, and were inhibited by photosynthetic inhibitors and uncouplers. The pH just outside the cell surface became about 5.6 (from 5.8) over most of the surface, with the alkaline bands marked by sharp peaks to 8.5–9.5. Lucas & Smith propose that the alkaline bands are produced by bicarbonate uptake together with hydroxide excretion, the process being regenerative, in that it can only begin at a slightly alkaline pH, and then it can drive the pH much higher. At this high pH, the visible banding with calcium carbonate can appear, and the presence of the bands makes continued alkali excretion more certain in those parts of the cell surface.

Since the light-stimulation of chloride influx only occurred in Smith's (1970) experiments below pH 9, and was half-maximum at pH 8, the influx of chloride is liable to be much lower in the alkaline bands than in the acid regions. This effect has been shown by Spear *et al.* (1969). It seems clear that chloride influx in charophyte cells in the light will be 'patchy' unless a reasonable concentration of buffer is present. The stimulating effect of basic buffers means that Good's zwitterion buffers are the natural choice for the task (Good *et al.*, 1966).

The banding phenomenon has not been reported for other giant algal cells.

Location of the products of carbon fixation

Little work has been done to fractionate cells to find what products of photosynthesis are dispersed or transported to which organelles and other compartments, and which are retained in the chloroplasts. The giant-celled algae seem rather suitable for this sort of inquiry. Analyses made by Hoagland & Davis (1929) and by Mercer & Mercer (1971), showed that only small amounts of carbon and nitrogen compounds were to be found in the vacuoles of *N. clavata* and *C. corallina* respectively. The tonoplast of *N. translucens* is impermeable to glucose (Smith 1967*a*). The same author (1967*b*) used ^{14}C to trace the products of carbon fixation, and separated vacuolar sap from cytoplasm. He found a reasonably constant proportion of 25–50% of the total fixed radioactivity to be in the vacuole, between 0.1 and 3 h. Its appearance there was rapid; the graph of vacuolar radioactivity against time was linear through the origin. The above observations seem less contradictory if it is noticed that the radioactivity in Smith's experiments does not represent a large amount of chemical substance. The labelled substances found in the vacuole were mainly sucrose and amino-acids with smaller amounts of sugar phosphates; the total was about 0.5 mmol m^{-2} after 3 h, or 2.5 mM in concentration in the vacuolar sap. This level may not be reached in cells photosynthesising under less 'luxurious' conditions; fixation rates of only 70 nmol m^{-2} s^{-1} would be

Plate 7 Accumulation of base (dark areas) and acid (light areas) outside two isolated internodal cells of *N. clavata*. The external medium contained phenol red indicator and mineral salts, including $KHCO_3$; the pH was 6.9. The cells had been in dark for 20 min and then light of about 2 W m^{-2} for 20 min before the photograph. The scale mark represents 10 mm. (From Spear, Barr & Barr, 1969.)

more common under natural pond conditions, and of this, some 20–40 nmol $m^{-2} s^{-1}$ respired away. Of the cytoplasmic radio-activity, 50% was found to be in starch, and sucrose, sugar phosphates and amino-acids accounted for most of the remainder.

Substances affecting fixation

As yet, only Smith & West (1969) and Larkum & Lilley (unpublished data) appear to have compared fixation in whole cells with that in chloroplasts isolated from the same species. The discussion below will largely be based on measurements on algal cells and angiosperm chloroplasts.

It is almost always found that CCCP inhibits carbon fixation, as would be expected if it uncouples photophosphorylation *in vivo*. On the other hand imidazole and ammonia, uncouplers of photophosphorylation in isolated chloroplasts, are without effect on carbon fixation in *N. translucens* and *C. corallina*. Gramicidin and nigericin uncouple very effectively *in vitro* and might be valuable in work with intact giant algal cells but have not yet been shown to have this effect in whole cells. Ouabain is without consistent effect on carbon fixation and is assumed to act directly on the transport ATPase. DCMU has the action on fixation expected from its known action on electron transport in PSII, namely, it causes about 50% inhibition at 0.3 μM. Because of the braking action of DCMU, non-cyclic phosphorylation coupled to net electron flow is inhibited and hence so is fixation.

Much less easy to explain is the action of phlorizin in stimulating fixation in *Hydrodictyon* in saturating light (Raven, 1968c) since this substance acts as an 'energy transfer inhibitor' with isolated chloroplasts. This means that phlorizin acts by stopping the phosphorylation step that leads to ATP, and consequently slows down coupled electron flow (fig. 9.1). Dio-9, a detergent-like substance of unknown structure, acts similarly *in vitro* to phlorizin but inhibits fixation mildly at higher concentrations than are needed for chloroplasts. Dio-9 may affect the plasmalemma directly (Smith & West, 1969). Also, its potency as a phosphorylation inhibitor varies with each batch so perhaps the variable effects on chloride transport are not surprising.

To return to phlorizin, Raven proposed that its action on the light-stimulated potassium influx coupled with sodium efflux in *H. africanum* was a selective inhibition of cyclic photophosphorylation. There is an element of cyclic reasoning in this:

$$\left\{\begin{array}{l} \text{K}^+/\text{Na}^+ \text{ transport} \\ \text{driven by far red} \\ \text{light can be} \\ \text{inhibited by CCCP} \end{array}\right\} \therefore \left\{\begin{array}{l} \text{Cyclic photophos-} \\ \text{phorylation exists} \\ \text{and produces the} \\ \text{ATP required for} \\ \text{K}^+/\text{Na}^+ \text{ transport} \end{array}\right\} \therefore \left\{\begin{array}{l} \text{If a substance affects} \\ \text{K}^+/\text{Na}^+ \text{ transport it is} \\ \text{also an uncoupler of} \\ \text{cyclic photophos-} \\ \text{phorylation} \end{array}\right\}$$

On the above reasoning we should have to include ouabain as an inhibitor of cyclic photophosphorylation which would not meet with anyone's approval.

Note that at no time has cyclic photophosphorylation been demonstrated in giant algal cells, let alone *Hydrodictyon*. The last stage of this reasoning has, in fact, been used by Raven (1970*a*) in work to test for the importance of cyclic and pseudo-cyclic photophosphorylation in carbon fixation. Thus, concentrations of desaspidin, 2, 4-DNP, antimycin A, salicylaldoxime and DSPD were chosen that had a maximal inhibiting effect on the 710 nm-light-requiring potassium influx (coupled to Na$^+$ efflux?) under anaerobic conditions. Then these substances were tested for inhibition of carbon fixation, during photosynthesis in air and in nitrogen. Desaspidin, 2, 4-DNP, etc. each had small inhibiting effects, the rates being 70–80% of controls.

To explain the stimulation of carbon fixation and simultaneous inhibition of the potassium influx by phlorizin requires more knowledge about PS in whole cells than is presently available. The absence of effect of Dio-9 on carbon fixation at concentrations where the cation pump is inhibited leads to the same impasse. Raven, MacRobbie & Neuman (1969) favour the following conclusion: 'that there is normally an excess of ATP production over what is required for CO$_2$ fixation, and that, for reasons of compartmentation or affinity for ATP, the extrachloroplastic reactions are the first to be inhibited when ATP synthesis is inhibited; this does not, however, explain why such differential effects are not seen with CCCP. . .'. It is becoming more commonly realised that many inhibitory substances have not only ambiguous roles in terms of the old dichotomy 'inhibitor'–'uncoupler', but also many sites of action. Even the considerable researches of Raven have not avoided all the logical difficulties of applying very simple models to quite complex systems. Similar mechanisms seem indeed to provide photosynthetic and respiratory and osmotic energy transfers, as Mitchell has led us to recognise: similar inhibitors interfere with all. Our attempts at the moment are not unlike trying to understand a complex, electronic, feedback control system by pouring over it specific solvents for copper, germanium or silicon, by turning up and down the mains supply voltage, and by similar gross methods. We should perhaps neither despair nor let the chain of hypotheses get too long.

CHAPTER 12

Protoplasmic streaming

The endless rotation of the protoplasm in charophyte plants is a sight just as beautiful, and nearly as mystifying, as it was when Corti first recorded its observation. In the intervening two hundred years many workers have been attracted by its simplicity, and baffled by that same simplicity. More complex in pattern, and less often investigated, the protoplasmic streaming in genera such as *Acetabularia* and *Bryopsis* may well be similar in mechanism to that in higher plant cells, and perhaps to that in the charophytes as well.

Streaming in the charophytes

Since words can hardly do justice to the phenomenon, any reader who has not observed streaming in a charophyte is urged to lay this book aside and to search his nearest river or Botany Department for some live specimens. The cytoplasm of each cell rotates steadily throughout its life, carrying the vacuolar contents with it, at a speed of some 40 to 120 μm s^{-1} near the cell wall. In the long cylindrical internodes the stream proceeds along the cell approximately axially in a hemi-cylindrical sheet, around the end of the cell, and back along the other half cylinder. The streams are separated, or rather their junction is marked, by an interruption in the packed rows of chloroplasts known as the white line, and by a slight inward ridge on the cell wall. Streaming is detected by the motion of the numerous cytoplasmic particles, by the motion of vacuolar particles in the same direction, and by the motion in the same direction, and at nearly the same speed, of bulges and irregularities in the tonoplast. This simple observation at once disposes of theories like that of Kavanau (1963) in which particles are supposed to be self-propelling and so to move in the opposite direction to the (invisible) ground plasm. The speed of streaming is measured from observations of one of the cytoplasmic particles, which have a range of speeds: the smallest particles generally have the highest and most consistent speed.

The stream is everywhere parallel to the rows of chloroplasts and, since the cell is twisted, the streaming is at an acute angle to the long axis, following a helical path of large pitch. The chloroplast rows are fixed, and so is the sense of the streaming – indeed the sense of the streaming and the morphology of the whole plant are correlated (fig. 12.1). In the

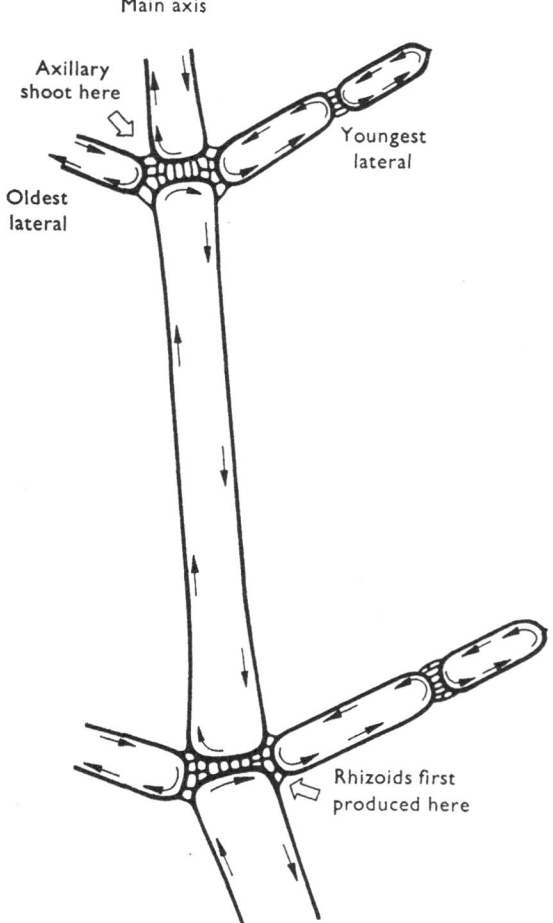

Main axis

Axillary
shoot here

Youngest
lateral

Oldest
lateral

Rhizoids first
produced here

Fig. 12.1 Part of a charophyte plant showing the directions of streaming of the cytoplasm in internodes of the axis and laterals.

163

node above an axial internode, the oldest lateral cell of the whorl appears above the ascending stream, the youngest above the descending stream. The axillary shoot also appears above the ascending stream, and at the basal node of an excised shoot, rhizoids appear first below the descending stream of the internode above. In the lateral cells the stream is always inwards (towards the axis) on the adaxial side and outwards on the abaxial side (Fritsch, 1965); at a lateral node this rule is preserved, so that there is a 'counter-current' effect in the streams on the opposite sides of the nodal septum.

The motive force

Streaming has often attracted those who delight in physical problems, from Ewart (1903) and his predecessors until today. At the microscopic level the physical situation has been most clearly defined by the beautiful work of Kamiya and his collaborators (e.g. Kamiya & Kuroda, 1956*b*, 1958*b*, 1965). They used normal internodes, and also lengths of internode which had been filled with protoplasm by gentle centrifugation and then tied off. In the latter preparation the distribution of streaming speeds was as shown in fig. 12.2*a*. The maximum speed occurs at the interface between solid cytoplasm containing chloroplasts, and the liquid endoplasm. It is clear by inspection that the motive force acts very close to this interface, at the site of maximum shear. The comparison with the velocity distribution in the myxomycete *Physarum* (fig. 12.2*b*) is both convincing and delightful, for *Physarum* is a good example of pressure-driven flow with passive shearing at the liquid–solid interface (Kamiya & Kuroda, 1958*a*).

At the microscopic level too, the connection between cyclosis and chloroplasts invites attention. That the stream runs parallel to the chloroplast rows has been noted; if chloroplasts are dislodged into the streaming endoplasm, they are seen to rotate (Jarosch, 1956; Fetzmann, 1958; Hayashi, 1964). This suggests that the mechanism producing the motive force is still attached to one side of the chloroplast. The chloroplast is not the only site of attachment of this mechanism as is clear from observation of streaming in the colourless rhizoid cells. Of the small cells in the node, too, some have fixed chloroplasts and others have chloroplasts which rotate with the

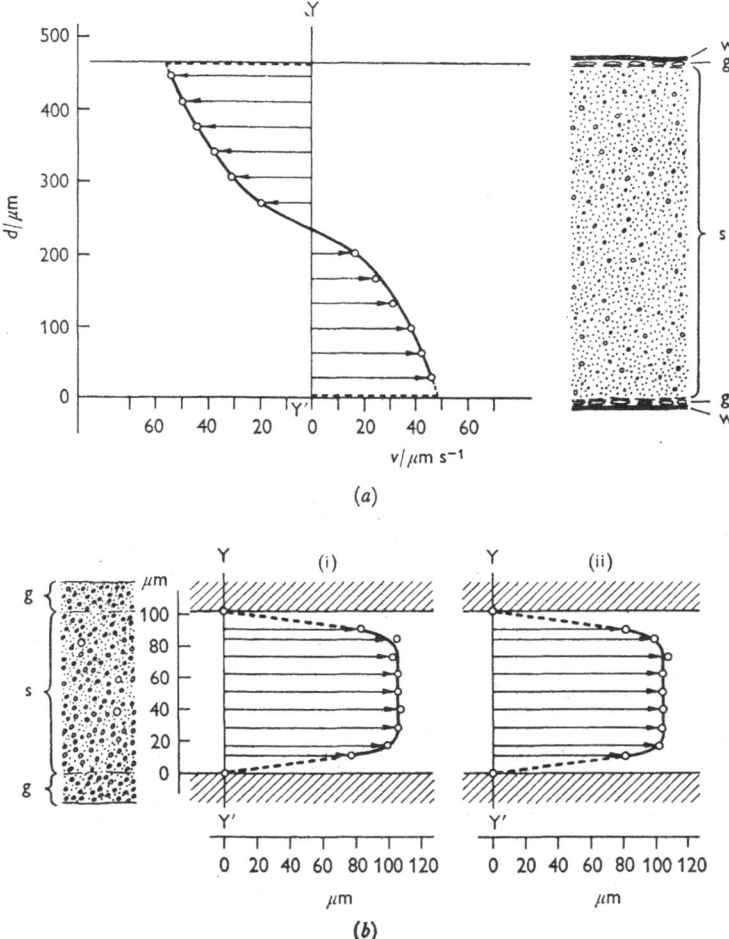

Fig. 12.2 (a) Velocity distribution (d) of cytoplasmic streaming in portion of a cell of *N. flexilis* in which the vacuole had been displaced by cytoplasm during centrifugation. On the right is a schematic transverse section, to the same scale as the left ordinate. w, cell wall; g, 'plasmagel' (containing chloroplasts embedded); s, 'plasmasol' (endoplasm, liquid cytoplasm). The rate of streaming in intact internodes was about 80 μm s^{-1}, temperature 28 °C (from Kamiya and Kuroda, 1956b). (b) The distribution of velocity of cytoplasmic streaming in a strand of the plasmodium of *Physarum polycephalum*. The lines joining YY' to the small circles represent distance travelled by granules in three seconds; (i) is streaming under natural conditions, (ii) streaming artificially produced when a pressure difference of 15 mm of water (148 Pa) was established across a strand 5 mm long. g, 'plasmagel', s, endoplasm (from Kamiya and Kuroda, 1958a).

cytoplasm. In green cells, whole areas of chloroplasts can be experimentally dislodged, and protoplasmic movement across these areas becomes obviously passive for the first 10–20 hours (Nichols, 1925; Hayashi, 1964). Organised, active streaming begins slowly in the clear regions, and later still they are colonised by chloroplasts again.

The mechanism that generates the motive force is thus replaceable, and becomes attached to whatever solid surface is available. Occasionally there has been observed vigorous streaming round and round a deeply inserted, sealed microelectrode.

The motive force itself has been measured by balancing it against centrifugal force in a centrifuge microscope (Kamiya & Kuroda, 1958b). Since the streaming motive force and the centrifugal force act on the fluid cytoplasm at different places, they cannot strictly be balanced against each other. For the motive force is at the interface, while the centrifugal force is distributed throughout the cytoplasm; when balanced they should cause the cytoplasm to shear. However, the narrow layer of cytoplasm behaves as a non-Newtonian fluid, of high viscosity for low shear stress (Kamiya & Kuroda, 1956b, 1965) and this no doubt helps to prevent shearing. Another problem is that the centrifugal force on the cytoplasm can only be calculated from the small density difference between cytoplasm and sap. However, the value of motive force obtained, 160 mN m^{-2}, is probably a fair approximation. Tazawa & Kishimoto (1968) have also measured the motive force as 170 mN m^{-2} by balancing it against the shearing force produced by vacuolar perfusion (see chapter 2).

These methods have been used to investigate two interesting and long-standing problems:

The effect of temperature on streaming rate

It is an old observation that the rate of streaming increases with temperature from 0 °C to about 30 °C (Dutrochet, 1837; Hörmann, 1898; Lambers, 1925; Romijn, 1931). A typical result (Umrath, 1934) is shown in fig. 12.3 for *Nitella mucronata*. It had never been clear whether the primary effect of temperature was on the viscosity of the cytoplasm or on the motive force. However, Hayashi (1960) showed that the motive force remains almost constant between 5 and 25 °C, so that in this range changes in the viscosity of the cytoplasm must be responsible for the variation of streaming rate with temperature (fig. 12.4).

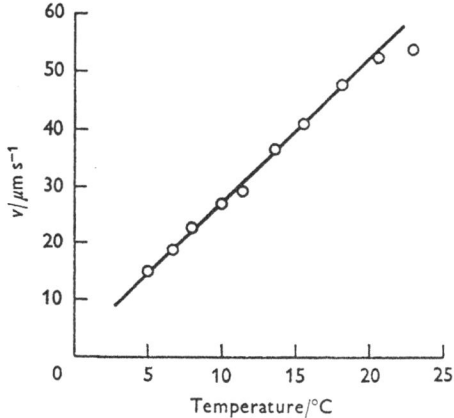

Fig. 12.3 The dependence of cytoplasmic streaming velocity on temperature in a cell of *N. mucronata* (from Umrath, 1934).

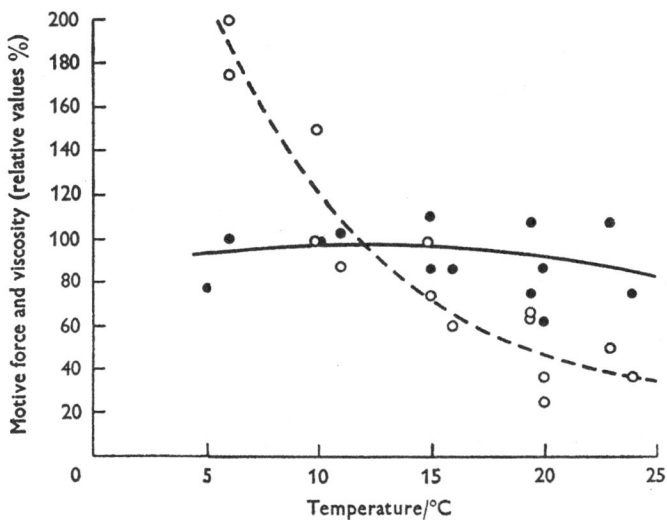

Fig. 12.4 The relation between motive force (●) and temperature, and between viscosity of the cytoplasm (○) and temperature, in cells of *C. corallina* (from Kamiya, 1962, after Hayashi, 1960).

Cessation of streaming produced by an action potential

It has also been known since 1898 that the passage of an action potential (see chapter 8) along a charophyte cell produces a temporary halt in the protoplasmic streaming (fig. 12.5a). The whole cell is normally affected, but if the action potential can be prevented from propagating, for

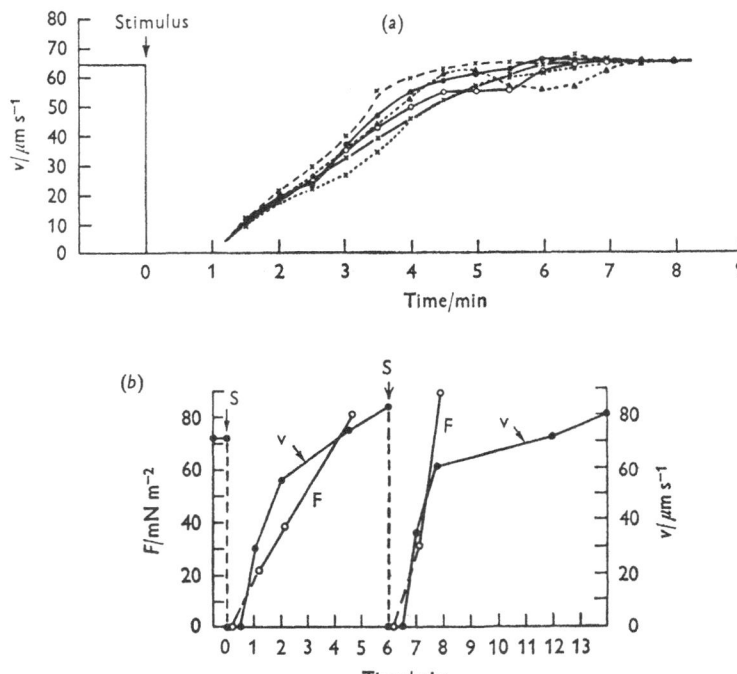

Fig. 12.5   (a) Cessation and subsequent recovery of streaming following a stimulus and AP at $t = 0$, in a cell of *Nitella*; temperature 19.5 °C (from Kishimoto & Akabori, 1959). (b) Two time courses of the recovery of the motive force (F) and streaming velocity (v) in a cell of *C. corallina* following an electrical stimulus (S) that produced an action potential (from Tazawa & Kishimoto, 1968).

example by an extracellular barrier of high electrical resistance, streaming stops only in that part of the cell along which propagation has occurred (Kishimoto & Akabori, 1959). The effect is cumulative, a train of several action potentials producing a longer pause in the streaming than a single one. It is not known what aspect of the excitation phenomenon affects streaming, but the methods already discussed can at least determine whether motive force or viscosity is affected. Gelfan (1928) attempted to answer this question by stimulating cells electrically, estimating viscosity by observing Brownian movement of cytoplasmic particles and by pricking the cell with a needle. He recorded that Brownian motion ceased 'for a brief time' upon stimulation; and that when a normally streaming cell was punctured, liquid contents escaped and dispersed, but when a cell was punctured just after stimulation, a coherent drop of protoplasm emerged slowly. He concluded that the effect was on viscosity. By contrast, Tazawa & Kishimoto (1968), using the perfusion balance method, found that there was normal (low) viscosity from several seconds after stimulation, and that the motive force fell to zero upon stimulation, recovering in parallel with the rate of streaming (fig. 12.5b).

As to the way in which excitation affects the motive forces the work of Barry (1968) is suggestive. He found in *Nitella axillaris* that the replacement of Ca^{2+} in the medium and cell wall with Mg^{2+} or Ba^{2+} left the cells excitable but blocked the stoppage of streaming that normally accompanies excitation. External Sr^{2+} behaves like Ca^{2+}. The simplest explanation would of course be that the action potential is accompanied by a small nett influx of divalent cations, among which Ca^{2+} and Sr^{2+} have an inhibitory effect on streaming. This explanation however lacks clear experimental support: Barry recalls one finding of Hope & Findlay (1964) in which a small influx of Ca^{2+} was detected during stimulation of cells poisoned with 2, 4-DNP.

Mechanical shock, bending and sudden drop in temperature (Hill, 1935) also cause a temporary halt in cyclosis. So far as is known, all such stimuli act on cyclosis by producing one or more action potentials.

Energy and its transduction

The source of energy for streaming is a problem that attracted early attention. Ewart (1903) found that streaming persisted for weeks in the dark and without oxygen. The easy inference from this, that ATP may be the immediate energy source, and that glycolytic pathways may provide sufficient ATP for the purpose, tends to be confirmed by more recent work, though

TABLE 12.1 *Effects of chemical agents on cyclosis in charophyte cells*

Substance	c/M	Effect on cyclosis	Reversible?	Supposed effect of substance	Remarks	Reference
PCMB	10^{-4}	I	No	Reacts with —SH groups	No APs	KA
PCMB	10^{-3}	I	By cysteine	,,	?	A
PCMBS	10^{-3}	IT		,,	Effect due to APs?	P
NEM	10^{-4}	I		,,		P
Hg(NO₃)₂	10^{-3}	I		,,	,,	P
CdCl₂	10^{-3}	O		,,	,,	P
Iodoacetamide	10^{-3}	I		,,	Effect due to APs?	P
Iodoacetate	10^{-4}	O		,,		H
IAA	10^{-4}	O		Plant growth hormone		P
ATP	10^{-2}	I		Donor of phosphate		SS
ATP	10^{-3}	S(15%)		,,	Penetrating species	SS
ATP		S		,,	not identified	VV
ATP	Any	O		,,		H
Ouabain	10^{-4}	O		I NaKMg ATPase		P
CN⁻	4.10^{-3}	O		I e.t.		P
CN⁻	10^{-3}	O		*I e.t.	Inhibits resp. O₂ by 50%	H
N₃⁻	10^{-4}	O		I e.t.		H
N₃⁻	10^{-3}	I	Yes	I e.t.		P
N₃⁻	10^{-2}	I		*I e.t.	Inhibits resp. O₂ by 60%	H
F⁻	10^{-2}	O		I e.t.		P
DCMU	10^{-5}	O		*I photo e.t.	Inhibits CO₂ fixation completely	P
Phlorizin	5.10^{-4}	O		*I oxyphos. and photophos.		**P**
Phlorizin	5.10^{-3}	IT		,,		P

Compound	Concentration	Effect	Reversible	Description	Notes	Ref
Arsenate	10^{-3}	O		I photophos and oxyphos.		P
Oligomycin	50^m	O		I oxyphos.		P
Dio-9	20^m	O		I photophos. and oxyphos.		P
DCCD	5×10^{-4}	I		I photophos, oxyphos. ox. e.t.		P
DNP	5×10^{-4}	I	Yes	U oxyphos.: S, ATPase		P
DNP	10^{-3}	I	Yes	"		H
DNP	2×10^{-5}	I in 6 days	Yes in light	"		BB
CCCP	5×10^{-6}	I	Yes	U photophos. and oxyphos.		P
Imidazole	10^{-3}	O		U photophos.		P
Imidazole	5×10^{-3}	I	Yes	"		P
L-ethionine	10^{-3}	O		Traps ATP	Axial internodes	P
L-ethionine	10^{-3}	I	?	" *	Terminal lateral internodes	P
Colchicine	10^{-3}	O		* Depolymerises microtubules		PH
Colchicine	10^{-2}	O		" *		P
Colchicine		O		" *		LS
8HOQ	10^{-3}	I	Yes	Colchicine analogue?		LS
Cytochalasin B	30^m	I	Yes	Dismantles some micro-filaments	Recovery occurs in presence of 10 mg l^{-1} of cycloheximide	B
Cytochalasin B	$1 - 50^m$	I	Yes	Inhibits micro-fibrillar movement		W

Key: I, inhibits; IT, inhibits temporarily; S, stimulates; O, no effect; U, uncouples; *, effect characterised on whole *Chara* cells; e.t., electron transports; m, mg l^{-1}.

References: A, Abe (1964); B, Bradley in Wessells *et al.* (1971); BB, Barr & Broyer (1966); PH, Pickett-Heaps (1967); H, Hayashi (1960); P, Polya (1968); KA, Kishimoto & Akabori (1959); SS, Sandan & Somura (1959); LS, Lucas & Smith, unpublished; VV, Vorobieva & Vorobiev (1965); W, Williamson (1972).

G

the picture is very confused. The obvious way to proceed is to bathe the cell for some time in a solution containing a chemical substance, one or more of whose effects on energy transfer systems has been determined *in vitro*. This approach has been used, and in table 12.1 are collected such data as have resulted from it. There are the same problems of interpretation here as have been illustrated in the discussion of energy sources for ion transport. There is also the specific problem in the charophytes that streaming almost always ceases if an action potential occurs.

Although cessation is brief for a single action potential, a train of them may halt cyclosis for a considerable time. This was pointed out by Polya (1968) who recorded the membrane potential during a number of his experiments. Although the membrane potential was generally affected by the reagents used, only with the sulphydryl group reagents PCMBS and NEM were there observed trains of action potentials that might have been responsible for the cessation of streaming. With PCMBS in particular cyclosis recovered soon after the action potentials ceased, suggesting that there was no direct effect. This is unfortunate, since other studies with PCMB (Sandan & Somura, 1959) had shown that the effect could be reversed by L-cysteine, and the conclusion that sulphydryl groups were involved in driving cyclosis would have been a welcome, if small, step forward.

With these reservations in mind, it is not surprising that no firm conclusion about energy supply can be reached from the inhibitor studies summarised in table 12.1. No doubt it can be hypothesised that ATP is involved.

Mechanism

Equally unsolved is the problem of the chemo-mechanical energy transducer. Microscopic observation reveals nothing to the casual observer: careful dark-field observations on drops of living endoplasm have shown motile fibres in various configurations (Jarosch, 1956; Kamitsubo, 1966). What appear to be linear aggregations of virtually invisible fibrils were seen, and these aggregations often formed polygonal figures. The polygons usually did not move, but their corners propagated around the figure at the rate of several revolutions per minute.

In the living cell, after gently centrifuging, linear bundles may occasionally be seen (Kamitsubo, 1966) and in one case one such fibre was seen to pinch off a polygonal figure. The suggestion is that these visible fibres are more or less abnormal forms of fibrillar streaming mechanism. The electron microscope makes visible small bundles of micro-filaments (Nagai & Rebhun, 1966) which appear to be our closest view of the energy transducer. The bundles are located on the surface of the chloroplasts facing the vacuole, spaced some 0.7 μm apart transversely. They consist of 50–100 micro-filaments of about 5 nm diameter. (The otherwise ubiquitous microtubules (of about 20 nm diameter) are not found here, though some are observed, running transversely, just inside the cell wall.)

Cytochalasin B inhibits streaming (table 12.1) without visibly altering the numbers of microfilaments. Williamson (1972) found similar numbers of visible fibrils in extruded protoplasmic drops in the presence and absence of cytochalasin B. So although the action of cytochalasin B is not yet known to be entirely specific for microfilaments, (Holtzer & Sanger, 1972) it is reasonable to conclude that bundles of microfilaments supply the motive force for streaming. The lack of action of colchicine (table 12.1) agrees with this idea: the inhibitory action of 8-hydroxyquinoline does not.

Everybody now expects that the molecular mechanism will turn out to bear some degree of resemblance to the mechanism of muscular contraction. Direct evidence for this connection is lacking, although extracts of charophyte cells have been shown to resemble actomyosin preparations (Poglazov, 1966).

Streaming in Acetabularia

The long cylindrical cells of *Acetabularia* have served for several investigations of protoplasmic streaming. In this cell streaming occurs along longitudinal striations in a cortical 'gel' layer of cytoplasm, various strands travelling up or down at about 2 μm s^{-1} (Kamiya & Kuroda 1966). Takata (1961) found marked but transient increases of the rate when ATP solutions were applied; glycerinated cells exhibited some form of mass movement of cytoplasm when ATP was applied. Kamiya & Kuroda (1966), removing cytoplasm from centrifuged cells, found a fibril connecting two drops of cytoplasm, along which particles moved jerkily, at different speeds, and in opposite directions. This is a very familiar situation in streaming in cytoplasmic strands in angiosperm cells – the mechanism may be fibrillar, as in *Nitella*, though we are hardly in a position to give this statement much meaning.

Streaming and teleology

Although we have been unable to make any useful suggestion about the function of the action potential in charophyte plants, we may hope to do so for cytoplasmic streaming. Its energy consumption of about 2 nW for a cell of surface area 10^{-4} m^2 is small compared to the $1-2$ μW available from respiration, even compared to the 200 nW used for chloride transport, but one would not wish it to be expended to no purpose.

While diffusion half-times are short over distances of 10 μm or so, they increase rapidly and become prohibitively long over distances of the order of 10 mm: effective chemical communication requires bulk movement. The uninucleate *Acetabularia* clearly needs such communication if the nucleus is to organise the growth of the cell. The multinucleate charophytes, having embarked on multicellularity and sizes approaching one metre, need chemical communication too. In these plants the rhizoids are supplied with photosynthetic products, and may be sites of starch storage (as in *Nitellopsis*). In the upper part of the plant there are growth correlations such as apical inhibition that suggest the movement of growth substances around the plant as well. The maximum growth rate casually observed for plants of *Chara corallina* is about 10 mm day^{-1}: this occurs in the sub-apical internode, which must therefore achieve a net gain of salt of about 10 pmol s^{-1}. If this were provided by a net influx across the plasmalemma of the elongating cell, it would need to be of the order of 300 nmol m^{-2} s^{-1}. In the small apical cells of *Nitella* studied by Green (1968), growth at constant turgor required a chloride entry to the growing cell of about 0.2 pmol s^{-1}; if this were a net influx across the plasmalemma the value would be about 60 nmol m^{-2} s^{-1}. These values are so high that it seems probable that the flux to the growing cell occurs via the plasmodesmata: certainly fast growth requires the presence of the basal internode. A flux of 300 nmol $m^{-2} s^{-1}$ across the plasmalemma of a *Chara* cell would require an energy input of the order of $1 \mu W$ in a cell of area $10^{-4} m^2$. This is comparable with the total available from respiration.

Indeed a corollary of streaming as a transport mechanism is the existence of intercellular transport at the node. There are plasmodesmata between neighbouring cells in *Nitella* (Fridavalsky, Lovas & Nagy, 1964; Spanswick & Costerton, 1967) and in *Chara* (Pickett-Heaps, 1967). These plasmodesmata, as Spanswick & Costerton, and Skierczynska (1968) both showed, have a high electric conductance, although the value Spanswick found was some 300 times lower than the value calculated for cylindrical pipes filled with material of the same conductivity as the sap. What structure is present in the lumen of the plasmodesma (plates 5a & b) is not clear.

Measurements have been made of the movement of labelled halide, from the solution surrounding one internodal cell to an adjacent internode, through the intervening node (Walker & Bostrom, 1973). In measurements over 3–6 h, in which the specific activity of the cytoplasm was approximately uniform, the measured flux across the node was about 25–40 pmol s^{-1}. This flux appears to be rate-limited not by movement across the node but by the rate at which halide is brought to the node by cytoplasmic streaming (T. E. Bostrom, unpublished data). These considerations indicate a content of chloride of about 150–250 μmol m^{-2} (say 25–40 mM). The difficulty of reconciling this with the data of fig. 6.7 is not yet resolved.

In an extensive qualitative study by means of whole-plant autoradiographs, Evrard & Chappel (1967) found that a number of labelled substances moved around the plant of *Chara vulgaris*, including sucrose, phosphate and the inhibitor simazin. As the simplest whole plant, *Chara* may be expected to be used for more such studies.

Appendix A

Notes on equations and their derivation

Equation 3.1 (p. 35)

$$\Psi_w = P - \Pi$$

This is derived from the expression for the chemical potential:

$\mu_w = RT \ln x_w + P \bar{V}_w + \mu_w^{(o)}$ (where x_w is the mole fraction of water)

by dividing by $\bar{V}_w$, assuming the latter to be constant.

Equation 4.6 (p. 52)

$$\mathcal{J}_v = L_P(\Delta P - \sigma \Delta \Pi)$$

is derived from the steady state thermodynamics of a system comprising membrane, water and a single solute, assuming perfect mixing. Interactions between solute and solvent are allowed for.

Equation 6.1 (p. 75)

$$\vec{\phi}_j = -\frac{z_j P_j F \psi_{co} c_j^0}{RT[1 - \exp(z_j F \psi_{co}/RT)]}$$

is derived from Goldman's (1943) equation for the net flux of ions j. In deriving this, Goldman assumed a linear potential gradient within the membrane, perfect mixing, and passive, independent diffusion of ions. To derive the equations for unidirectional fluxes, no further assumptions are necessary.

Equation 7.1 (p. 86)

$$\psi_{10} = (RT/F) \ln \{([K^+]_0 + \alpha[Na^+]_0)/([K^+]_1 + \alpha[Na^+]_1)\}$$

was derived by Hodgkin & Katz (1949), assuming the membrane to be permeable solely to K^+ and Na^+, and assuming constant permeabilities (related to ion mobilities and partition coefficients).

Equation 7.2 (p. 89)

$$\psi_{co} = \frac{RT}{F} \ln \frac{[K^+]_o + \alpha[Na^+]_o + \gamma[Cl^-]_c}{[K^+]_c + \alpha[Na^+]_c + \gamma[Cl^-]_o}$$

may be derived from Goldman's flux equations (6.1) by summing the net fluxes of K^+, Na^+ and Cl^- for zero external current, assuming a membrane with a linear potential gradient within it, with constant permeability to K^+, Na^+ and Cl^- only, passive independent diffusion and perfect mixing in the solution phases. It further assumes the phase boundary p.d.s are negligible; in this it differs from (7.1).

Equation 7.3 (p. 91)

$$\psi_j = (RT/z_jF) \ln (c_j^o/c_j^c)$$

This equation can be derived assuming electrochemical equilibrium and writing the usual expression for the electrochemical potential of the ion species:

$$\bar{\mu}_j = \mu_j^{(o)} + P\bar{V}_j + RT \ln a_j + z_jF\psi$$

In this particular version of the Nernst equation, activity coefficients are supposed equal in phases o and c.

Equation 7.6 (p. 92)

$$G_j = \Delta \mathcal{J}_j/\Delta \psi_M = \frac{P_j c_j^o F^2 \psi_M [\mathrm{I} - \exp(F\Delta \psi_M/RT)]}{RT\Delta \psi_M [\mathrm{I} - \exp(F\psi_M/RT)]}$$

can be derived from Goldman's equation for net flux by setting

$$\psi = \psi_M + \Delta \psi_M.$$

Equation 7.7 (p. 97)

$$g_j = z_j^2 F^2 \phi_{joc}/RT$$

comes from the definition of g_j (7.8), together with the assumption of electrochemical equilibrium. The influx and efflux, which are equal, must be assumed to be independent and passive.

Equation 7.9 (p. 99)

$$\ln(\phi_{joc}/\phi_{jco}) = n[\ln(c_o/c_c) - z_jF\psi_{co}/RT]$$

is a version of the Ussing flux-ratio equation [cf. 9.2] with an empirical factor n. Empirical in its origin, this equation is a reduced form of equation 9.1.

Equation 9.1 (p. 120)

$$RT \ln(\phi_{jab}/\phi_{jba}) = (\mathcal{R}^*/\mathcal{R})(\bar{\mu}_a - \bar{\mu}_b) + \sum_{k=1}^{k=m} \mathcal{J}_k \int_0^\delta (r_{jk}/r_{jj}) \, dx$$
$$+ \mathcal{J}_n \int_0^\delta (r_{jn}/r_{jj}) \, dx$$

The flux-ratio equation of Kedem & Essig (1965), based on steady state thermodynamics of a membrane system. It is assumed that a uniform membrane separates perfectly mixed components; components 1 to $n-1$ may move across the membrane, and all possible interactions i–j are allowed for including interactions between particles of the same species. In addition there may be an interaction between metabolism and each mobile component. In this equation $\mathcal{R}^*/\mathcal{R}$ is a positive constant $\geqslant 1$.

Equation 9.2 (p. 122)

$$RT \ln(\phi_{jab}/\phi_{jba}) = \bar{\mu}_a - \bar{\mu}_b$$

This is the simplest form of the flux-ratio equation, which assumes that the unidirectional fluxes of the diffusing species are passive and independent. This equation is a simplified version of that derived by Ussing (1949), omitting solute–solvent interactions which he allowed for. It is of course also a simplified version of 9.1.

Appendix B

Rates of energy conversion in Chara & Griffithsia

For vacuolated cells it is convenient to calculate rates of energy conversion (power) per unit area of cell surface, i.e. in W m^{-2}.

Chara corallina

(a) Photosynthesis (table 11.1). We calculate the rate at which energy is stored in glucose, assuming a photosynthetic quotient of 1.0:

$$\frac{1}{A} \cdot \frac{dG}{dt} = \mathcal{J}_{CO_2} \times \text{molar free energy of glucose oxidation}/6$$

$$= 400 \times 10^{-9} \times 2.89 \times 10^6/6 \ \text{W m}^{-2}$$

$$= 190 \ \text{mW m}^{-2}$$

(b) Respiration (unpublished data of R. J. Reid). We calculate the rate of passage of energy through ATP:

$$\frac{1}{A} \cdot \frac{dG}{dt} = \mathcal{J}_{O_2} \times \text{molar free energy of ATP hydrolysis} \times 38/6$$

$$= 140 \times 10^{-9} \times 33 \times 10^3 \times 38/6 \ \text{W m}^{-2}$$

$$= 29 \ \text{mW m}^{-2}$$

(c) Cytoplasmic streaming (chapter 12). We calculate the dissipation:

$$\Phi = v \times \text{Force}$$

$$= 100 \times 10^{-6} \times 0.16 \ \text{W m}^{-2}$$

$$= 16 \ \mu\text{W m}^{-2}$$

(d) Ion transport (chapter 6, especially fig. 6.7). The total dissipation due to ion movement cannot easily be calculated, but we can as an approximation calculate the rate of energy storage by various active transport systems: in the steady state this would equal the dissipation by passive fluxes. An allowance which is not known needs to be made for the 'inefficiency' of the pump. In all cases below (except that of hydrogen) the flux used in the calculation is the labellable (one-way) flux, not the net flux. Both exchange diffusion and passive diffusion in the direction of active transport are thus ignored. We write:

$$\frac{1}{A} \cdot \frac{dG}{dt} = \mathcal{J}_1 \cdot \Delta\bar{\mu}_1 = \mathcal{J}_1 \left(RT \ln \frac{c_1}{c_2} + z_1 F \psi_{12} \right)$$

Assuming the values of fig. 6.7, that $\psi_{co} = -170\,\mathrm{mV}$, and $\psi_{vc} = +15\,\mathrm{mV}$, and that the cytoplasmic phase is 6 μm thick

Chloride, plasmalemma, light:

$$\frac{1}{A}\cdot\frac{dG}{dt} = 23\times 10^{-9}(2.43\times 10^3\times\ln 5 + 96480\times 0.17)\ \mathrm{W\,m^{-2}}$$

$$= 470\ \mu\mathrm{W\,m^{-2}}$$

Since $\Delta\bar{\mu}_{Cl}$ is 20.3 kJ mol^{-1}, on the basis of the present assumptions the stoichiometry of the pump would have to be 1 ATP: 1 Cl$^-$, and the total energy expenditure would be $\approx 760\ \mu\mathrm{W\,m^{-2}}$.

Chloride, plasmalemma, dark:

$$\frac{1}{A}\cdot\frac{dG}{dt} = 35\ \mu\mathrm{W\,m^{-2}}$$

(as before, total energy expenditure $\approx 56\ \mu\mathrm{W\,m^{-2}}$)

Chloride, tonoplast, light:

$$\frac{1}{A}\cdot\frac{dG}{dt} = 1.3\ \mathrm{mW\,m^{-2}}$$

Chloride, tonoplast, dark:

$$\frac{1}{A}\cdot\frac{dG}{dt} = 850\ \mu\mathrm{W\,m^{-2}}$$

Sodium, plasmalemma:

assuming $[\mathrm{Na}]_c = 20$ mM and that there is no coupling to K$^+$,

$$\frac{1}{A}\cdot\frac{dG}{dt} = 4\times 10^{-9}(2.43\times 10^3\times\ln 20 + 96480\times 0.17)\ \mathrm{W\,m^{-2}}$$

$$= 95\ \mu\mathrm{W\,m^{-2}}$$

as before, $\Delta\bar{\mu}_{Na} = 23.7$ kJ mol^{-1} so that total expended energy

$$\approx 130\ \mu\mathrm{W\,m^{-2}}$$

Sodium, tonoplast: if $\phi_{cv} \approx 200$ nmol m^{-2} s^{-1}, $[\mathrm{Na}]_c \approx 20$ mM:

$$\frac{1}{A}\cdot\frac{dG}{dt} = 820\ \mu\mathrm{W\,m^{-2}}$$

Hydrogen, plasmalemma: if pH$_c \approx 7$, pH$_0 \approx 5.5$, $\mathcal{J}_H \approx 200$ nmol m^{-2} s^{-1}

$$\frac{1}{A}\cdot\frac{dG}{dt} = 5\ \mathrm{mW\,m^{-2}}$$

Griffithsia

(a) Photosynthesis (Lilley & Hope, 1971*b*).

$$\frac{1}{A}\cdot\frac{dG}{dt} = 170\ \mathrm{mW\,m^{-2}}$$

179

G*

(b) Respiration.

$$\frac{1}{A} \cdot \frac{dG}{dt} = 26 \text{ mW m}^{-2}$$

(c) Ion transport.

Chloride, plasmalemma, light, if $[Cl]_c$ is in equilibrium with $[Cl]_v$:

$$\frac{1}{A} \cdot \frac{dG}{dt} = 250 \,\mu\text{W m}^{-2}$$

Chloride, plasmalemma, dark:

$$\frac{1}{A} \cdot \frac{dG}{dt} = 55 \,\mu\text{W m}^{-2}$$

References

Abe, S. (1964) The effect of p-chloromercuribenzoate on rotational protoplasmic streaming in plant cells. *Protoplasma* **58**, 483.

Adrianov, V. K., Kurella, G. A. & Litvin, F. F. (1965) Change in potential of cells of the algae *Nitella* exposed to light and the connexion of this effect with photosynthesis. *Biophysics* **10**, 588.

Adrianov, V. K., Vorobieva, I. A. & Kurella, G. A. (1968) Investigation of the resting potential of *Nitella* cells. 2. Effect of pH of medium on resting potential of *Nitella* cells. *Biophysics* **13**, 396.

Aikman, D. P. & Dainty, J. (1966) Ionic relations of *Valonia ventricosa*. In *Some contemporary studies in marine science*, ed. H. Barnes, p. 37 George Allen & Unwin, London.

Andreoli, T. E., Tieffenberg, M. & Tosteson, D. C. (1967) The effect of valinomycin on the ionic permeability of thin lipid membranes. *J. gen. Physiol.* **50**, 2527.

Arens, K. (1936) Physiologisch polarisierter Massenaustauch und Photosynthese bei submersen Wasserpflanzen. II. Die $Ca(HCO_3)_2$ – Assimilation. *Jahrb. wiss. Bot.* **83**, 513.

(1939) Physiologische Multipolarität der Zelle von *Nitella* während der Photosynthese. *Protoplasma* **33**, 295.

Atkinson, M. R. & Polya, G. (1967) Salt-stimulated adenosine triphosphatases from carrot, beet and *Chara australis*. *Aust. J. biol. Sci.* **20**, 1069.

Barr, C. E. (1965) Na and K fluxes in *Nitella clavata*. *J. gen. Physiol.* **49**, 181.

Barr, C. E. & Broyer, T. C. (1964) Effect of light on sodium influx, membrane potential, and protoplasmic streaming in *Nitella*. *Pl. Physiol., Lancaster* **39**, 48.

(1966) Energy balancing in *Nitella* cells treated with DNP. *Science, Washington* **151**, 1245.

Barry, P. H. (1967) Investigation of the movement of water and ions in plant cell membranes. Ph.D. thesis, University of Sydney.

(1970) Volume flows and pressure changes during an action potential in cells of *Chara australis*. I. Experimental results. *J. memb. Biol.* **3**, 313. II. Theoretical considerations. *J. memb. Biol.* **3**, 335.

Barry, P. H. & Hope, A. B. (1969a) Electro-osmosis in membranes: Effects of unstirred layers and transport numbers. I. Theory. *Biophys. J.* **9**, 700.

(1969b) Electro-osmosis in membranes: Effects of unstirred layers and transport numbers. II. Experimental. *Biophys. J.* **9**, 729.

(1969c) Electro-osmosis in *Chara* and *Nitella* cells. *Biochim. biophys. Acta* **193**, 124.

Barry, W. H. (1968) Coupling of excitation and cessation of cyclosis in *Nitella*: role of divalent cations. *J. cell Physiol.* **72**, 153.

Barton, R. (1965a) An unusual organelle in the peripheral cytoplasm of *Chara* cells. *Nature* **205**, 201.

(1965b) Electron microscope studies on surface activity in cells of *Chara*. *Planta*, Berlin **66**, 95.

(1968) Autoradiographic studies on wall formation in *Chara*. *Planta*, Berlin **82**, 302-6.

Bisalputra, T. T. (1974) In *Physiology and Biochemistry of Algae*, ed. G. E. Fogg (in press).

Blinks, L. R. (1929) Protoplasmic potentials in *Halicystis*. *J. gen. Physiol.* **13**, 223.

(1930a) The direct current resistance of *Valonia*. *J. gen. Physiol.* **13**, 361.

(1930b) The direct current resistance of *Nitella*. *J. gen. Physiol.* **13**, 495.

(1935) Protoplasmic potentials in *Halicystis*. IV. Vacuolar perfusion with artificial sap and sea water. *J. gen. Physiol.* **18**, 409.

(1949) The source of the bioelectric potentials in large plant cells. *Proc. natn. Acad. Sci.* **35**, 566.

Blinks, L. R. Harris, E. S. & Osterhout, W. J. V. (1929) Studies on stimulation in *Nitella*. *Proc. Soc. exp. Biol. Med.* **26**, 836.

Blinks, L. R. & Jacques, A. G. (1929) The cell sap of *Halicystis*. *J. gen. Physiol.* **13**, 733.

Blount, R. W. & Levedahl, B. H. (1960) Active sodium and chloride transport in the single-celled marine alga, *Halicystis ovalis*. *Acta physiol. Scand* **49**, 1.

Bonting, S. L. (1970) Sodium-potassium activated adenosine-triphosphatase and cation transport. In *Membranes and ion transport*, vol. I, ed. E. E. Bittar, p. 257, Wiley-Interscience, London and New York.

Briggs, G. E. (1962) Membrane potential differences in *Chara australis*. *Proc. R. Soc. B* **156**, 573.

(1967) Electro-osmosis in *Nitella*. *Proc. R. Soc. B* **168**, 22.

Briggs, G. E., Hope, A. B. & Robertson, R. N. (1961) *Electrolytes and plant cells*, Blackwell, Oxford.

Brooks, M. M. (1922) Penetration of cations into living cells. *J. gen. Physiol.* **4**, 347.

Brooks, S. C. (1917) A new method of studying permeability. *Bot. Gaz.* **64**, 306.

Brooks, S. C. & Gelfan, S. (1928) Bioelectric potentials in *Nitella*. *Protoplasma* **5**, 86.

Bugarsky, St. & Tangl, F. (1898) Physikalisch-chemische Untersuchungen über die Molekularen Concentrations-verhaltnisse des Blutserums. *Arch. ges. Physiol.* **72**, 531.

Cass, A. & Finkelstein, A. (1967) The water permeability of thin lipid membranes. *J. gen. Physiol.* **50**, 1765.

Chambers, R. (1922) A micro injection study on the permeability of the starfish egg. *J. gen. Physiol.* **5**, 189.

Cohn, F. (1871) Ober des Gefrieren der Zellen von *Nitella syncarpa*. *Bot. Zeitg.* **29**, 723.

Cole, K. S. & Curtis, H. J. (1938) Electric impedance of *Nitella* during activity. *J. gen. Physiol.* **22**, 37.

Collander, R. (1930) Permeabilitätsstudien an *Chara ceratophylla*. I. Die normale Zusammensetzung des Zellsaftes. *Acta bot. fenn.* **6**, 1.

(1949*a*) The permeability of plant protoplasts to small molecules. *Physiologia Pl.* **2**, 300.

(1949*b*) Die Verteilung organischer Verbindungen zwischen Äther und Wasser. *Acta chem. scand.* **3**, 717.

(1950) The distribution of organic compounds between *iso*-butanol and water. *Acta chem. scand.* **4**, 1085.

(1954) The permeability of *Nitella* cells to nonelectrolytes. *Physiologia Pl.* **7**, 420.

Collander, R. & Bärlund, H. (1933) Permeabilitätsstudien an *Chara ceratophylla*. II. Die permeabilität für nichtelektrolytes. *Acta. bot. fenn.* **11**, 1.

Corti, B. (1774) *Osservazioni microscopiche sulla tremella e sulla circolazione del fluido in una pianta aquajuola*, Lucca.

(1776) Sur la circulation d'un fluide, de'couverti en diverses plantes. *Rosier obs. sur la Physique, sur l'Histoire Nat.* **8**, 232.

Coster, H. G. L. (1964) Intracellular chloride and its role in the electrical behaviour of the membranes of *Chara australis*. M.Sc. Thesis, University of Sydney, Sydney.

(1965) A quantitative analysis of the voltage–current relationships of fixed charge membranes and the associated property of 'punch-through'. *Biophys. J.* **5**, 671.

(1966) Chloride in cells of *Chara australis*. *Aust. J. biol. Sci.* **19**, 545.

(1969) The role of pH in the punch-through effect in the electrical characteristics of *Chara australis*. *Aust. J. biol. Sci.* **22**, 365.

(1973*a*) The double fixed charge membrane. Low frequency dielectric dispersion. *Biophys. J.* **13**, 118.

(1973*b*) The double fixed charge membrane. Solution–membrane ion partition effects and membrane potentials. *Biophys. J.* **13**, 133.

Coster, H. G. L. & George, E. P. (1968) A thermodynamic analysis of fluxes and flux-ratios in biological membranes. *Biophys. J.* **8**, 457.

Coster, H. G. L. & Hope, A. B. (1968) Ionic relations of *Chara australis*. XI. Chloride fluxes. *Aust. J. biol. Sci.* **21**, 243.

Coster, H. G. L., Syriatowicz, J. C. & Vorobiev, L. N. (1968) Cytoplasmic ion exchange during rest and excitation in *Chara australis*. *Aust. J. biol. Sci.* **21**, 1069.

Costerton, J. W. F. & MacRobbie, E. A. C. (1970) Ultrastructure of *Nitella translucens* in relation to ion transport. *J. exp. Bot.* **21**, 535.

Crawley, J. C. W. (1965) A cytoplasmic organelle associated with the cell walls of *Chara* and *Nitella*. *Nature*, London **205**, 200.

Crozier, W. J. (1919) Intracellular acidity in Valonia. *J. gen. Physiol.* **1**, 581.

Curtis, H. J. & Cole, K. S. (1937) Transverse electric impedence of *Nitella*. *J. gen. Physiol.* **21**, 189.

Dainty, J. (1962) Ion transport and electric potentials in plant cells. *Ann. Rev. Pl. Physiol.* **13**, 379.

(1963*a*) Water relations of plant cells. *Adv. bot. Res.* **1**, 279.

(1963*b*) The polar permeability of plant cell membranes to water. *Protoplasma* **62**, 220.

Dainty, J., Croghan, P. C. & Fensom, D. S. (1963) Electro-osmosis with some applications to plant physiology. *Can. J. Bot.* **41**, 953.

Dainty, J. & Ginzburg, B. Z. (1963) Irreversible thermodynamics and frictional models of membrane processes, with particular reference to the cell membrane. *J. theoret. Biol.* **5**, 256.

(1964a) The measurement of hydraulic conductivity (osmotic permeability to water) of internodal characean cells by means of transcellular osmosis. *Biochim. biophys. Acta* **79**, 102.

(1964b) The permeability of the cell membranes of *Nitella translucens* to urea and the effect of high concentrations of sucrose on this permeability. *Biochim. biophys. Acta* **79**, 112.

(1964c) The permeability of the protoplasts of *Chara australis* and *Nitella translucens* to methanol, ethanol and iso-propanol. *Biochim. biophys. Acta* **79**, 122.

(1964d) The reflection coefficient of plant cell membranes for certain solutes. *Biochim. biophys. Acta* **79**, 129.

Dainty, J. & Hope, A. B. (1959a) The water permeability of cells of *Chara australis* R. Br. *Aust. J. biol Sci.* **12**, 136.

(1959b) Ionic relations of cells of *Chara australis*. I. Ion exchange in the cell wall. *Aust. J. biol. Sci.* **12**, 395.

(1961) The electric double layer and the Donnan equilibrium in relation to plant cell walls. *Aust. J. biol. Sci.* **14**, 541.

Dainty, J., Hope, A. B. & Denby, C. (1960) Ionic relations of cells of *Chara australis*. II. The indiffusible anions of the cell wall. *Aust. J. biol. Sci.* **13**, 267.

Damon, E. B. (1929) Dissimilarity of inner and outer protoplasmic surfaces in Valonia II. *J. gen. Physiol.* **13**, 207.

Davies, J. T. & Rideal, E. K. (1961) *Interfacial phenomena*, Academic Press, New York and London.

Diamond, J. M. & Solomon, A. K. (1959) Intracellular potassium compartments in *Nitella axillaris*. *J. gen. Physiol.* **42**, 1105.

Diamond, J. M. & Wright, E. M. (1969) Biological membranes: the physical basis of ion and nonelectrolyte selectivity. *Ann. Rev. Physiol.* **31**, 581.

Dick, D. A. T. (1966) *Cell water*, Butterworth, London.

Dodd, W. A., Pitman, M. G. & West, K. R. (1966) Sodium and potassium transport in the marine alga *Chaetomorpha darwinii*. *Aust. J. biol. Sci.* **19**, 341.

Dutrochet, H. (1837) Observations sur la circulation des fluides chez le *Chara fragilis*. 7. Influence de la temperature sur la circulation du *Chara*. *Annls. Sci. nat. Ser.* 2, **9**, 24.

Evrard, T. O. & Chappel, W. E. (1967) Translocation of growth regulators in *Chara vulgaris*, a nonvascular aquatic plant found in Virginia's waters. *Water Resour. Res. Cent. Bull.*, Blacksburg, Va **5**, 1.

Ewart, A. J. (1903) *On the physics and physiology of protoplasmic streaming in plants*, Oxford University Press, Oxford.

Fensom, D. & Dainty, J. (1963) Electro-osmosis in *Nitella*. *Can. J. Bot.* **41**, 685.

Fetzmann, E. L. (1958) Uber rotierende eigen Bewegung der Zellkerne und Plastiden bei *Chara foetida*. *Protoplasma* **49**, 549.

Findlay, G. P. (1959) Studies of action potentials in the vacuole and cytoplasm of Nitella. *Aust. J. biol. Sci.* **12**, 412.

(1961) Voltage clamp experiments with *Nitella*. *Nature, London* **191**, 812.

(1962) Calcium ions and the action potential in *Nitella*. *Aust. J. biol. Sci.* **15**, 69.

(1964) Ionic relations of cells of *Chara australis*. VIII. Membrane currents during voltage clamp. *Aust. J. biol. Sci.* **17**, 388.

(1970) Membrane electrical behaviour in *Nitellopsis obtusa*. *Aust. J. biol. Sci.* **23**, 1033.

Findlay, G. P. & Hope, A. B. (1964) Ionic relations of cells of *Chara australis*. VII. The separate electrical characteristics of the plasmalemma and tonoplast. *Aust. J. biol. Sci.* **17**, 62.

Findlay, G. P., Hope, A. B., Pitman, M. G., Smith, F. A. & Walker, N. A. (1969) Ionic fluxes in cells of *Chara corallina*. *Biochim. biophys. Acta* **183**, 565.

Findlay, G. P., Hope, A. B., Pitman, M. G., Smith, F. A. & Walker, N. A. (1971) Ionic relations of marine algae. III. *Chaetomorpha*: membrane electrical properties and chloride fluxes. *Aust. J. biol. Sci.* **24**, 731.

Findlay, G. P., Hope, A. B. & Walker, N. A. (1971) Quantization of a flux ratio in charophytes? *Biochim. biophys. Acta* **233**, 155.

Findlay, G. P., Hope, A. B. & Williams, E. J. (1969) Ionic relations of marine algae. I. *Griffithsia*: membrane electrical properties. *Aust. J. biol. Sci.* **22**, 1163.

(1970) Ionic relations of marine algae. II. *Griffithsia*: ionic fluxes. *Aust. J. biol. Sci.* **23**, 323.

Finkelstein, A. & Cass, A. (1968) Permeability and electrical properties of thin lipid membranes. *J. gen. Physiol.* **52**, 145s.

Fischer, E. (1915) *Oedema and Nephritis*, New York.

Forsberg, C. (1965) Nutritional studies of *Chara* in axenic cultures. *Physiol. Pl.* **18**, 275.

Fraenckel, P. (1904) Uber die Bestimmung des Blutkörperchenvolumens aus der elektrischen Leitfähigkeit. *Zeitschr. klin. Med.* **52**, 470.

Fridavalsky, L., Lovas, B. & Nagy, T. (1964) Light and electron microscopic investigations of the cell wall of the *Charophyceae*. *Bot. Kozlemenyek* **51**, 211.

Fritsch, F. E. (1965) *The structure and reproduction of the algae.* Cambridge Univ. Press, Cambridge.

Fujita, M. (1962) Electrophysiological studies of *Nitella* cells, with special reference to the electric resistance. *Pl. Cell Physiol.* **3**, 229.

Gaffey, C. T. & Mullins, L. J. (1958) Ion fluxes during the action potential in *Chara*. *J. Physiol.* **144**, 505.

Gelfan, S. (1927) Conductivity of protoplasm (*in situ*). *Univ. Cal. Publ. Zool.* **29**, 453.

(1928) The electrical conductivity of protoplasm. *Protoplasma* **4**, 192.

Gicklhorn, J. & Umrath, K. (1928) Messungen elektrische Potentiale pflanzliche Gewebe und einzelner Zelle. *Protoplasma* **4**, 228.

Goldman, D. E. (1943) Potential, impedance and rectification in membranes. *J. gen. Physiol.* **27**, 37.

Good, N. E., Winget, D., Winter, W., Connolly, T. N., Izawa, S. & Singh, R. M. M. (1966) Hydrogen ion buffers for biological research. *Biochemistry* 5, 467.

Gradmann, D. (1970) Einfluss von Licht, Temperatur und Aussenmedium auf das elektrische Verhalten von *Acetabularia crenulata*. *Planta, Berlin* 93, 323.

Green, P. B. (1968) Growth physics in *Nitella*: a method for continuous *in vivo* analysis of extensibility based on a micromanometer technique for turgor pressure. *Plant Physiol.* 43, 1169.

Green, P. B. & Stanton, F. W. (1967) Turgor pressure: direct manometric measurement in single cells of *Nitella*. *Science, Washington* 155, 1675.

Gunning, B. S. (1965) The greening process in plastids. I. The structure of the prolamellar body. *Protoplasma* 60, 111.

Gutknecht, J. (1966) Sodium, potassium and chloride transport and membrane potentials in *Valonia ventricosa*. *Biol. Bull.* 130, 331.

(1967a) Ion fluxes and short-circuit current in internally perfused cells of *Valonia ventricosa*. *J. gen. Physiol.* 50, 1821.

(1967b) Membranes of *Valonia ventricosa*: apparent absence of water-filled pores. *Science, Washington* 158, 787.

(1968a) Salt transport in *Valonia*: inhibition of potassium uptake by small hydrostatic pressures. *Science, Washington* 160, 68.

(1968b) Permeability of *Valonia* to water and to solutes: apparent absence of aqueous membrane pores. *Biochim. biophys. Acta* 163, 20.

Haake, O. (1892) Uber die Ursachen electrischer Strome in Pflanzen. *Flora* 75, 455.

Hansen, A. (1893) Uber Stoffbildung bei den Meeresalgen. *Mitt. zool. Stn. Neapel* 11, 255.

Hansen, U-P. (1971) The frequency response of the action of light on the membrane potential of *Nitella*. *Biophysik* 7, 223.

Hansen, U-P. & Gradmann, D. (1971) The action of sinusoidally modulated light on the membrane potential of *Acetabularia*. *Pl. Cell Physiol.* 12. 335.

Hayashi, T. (1960) *Sci. Papers Coll. Gen. Educ. Univ. Tokyo* 10, 245.

(1964) Role of the cortical gel layer in cytoplasmic streaming. In *Primitive motile systems in cell biology*, p. 19, eds R. D. Allen & N. Kamiya, Academic Press, New York.

Heber, U. & Krause, G. H. (1971) Transfer of carbon, phosphate energy and reducing equivalents across the chloroplast envelope. In *Photosynthesis and photorespiration*, eds M. D. Hatch, C. B. Osmond & R. O. Slatyer, p. 218, Wiley-Interscience, New York, London, Sydney, Toronto.

Heber, U. & Santarius, K. A. (1970) Direct and indirect transfer of ATP and ADP across the chloroplast envelope. *Z. Naturforsch.* 25b, 718.

Hill, S. E. (1935) Stimulation by cold in *Nitella*. *J. gen. Physiol.* 18, 357.

Hoagland, D. R. & Davis, A. R. (1923) The composition of the cell sap of the plant in relation to the absorption of ions. *J. gen. Physiol.* 5, 629.

(1929) The intake and accumulation of electrolytes by plant cells. *Protoplasma* 6, 610.

Hodgkin, A. L. (1951) The ionic basis of electrical activity in nerve and muscle. *Biol. Rev.* 26, 339.

Hodgkin, A. L. & Huxley, A. F. (1952) A quantitative description of membrane current and its application to conduction and excitation in nerve. *J. physiol.* **117**, 500.

Hodgkin, A. L. & Katz, B. (1949) The effect of sodium ions on the electrical activity of the giant axon of the squid. *J. physiol.* **108**, 37.

Hodgkin, A. L. & Keynes, R. D. (1955) The potassium permeability of a giant nerve fibre. *J. physiol.* **128**, 61.

Hogg, J., Williams, E. J. & Johnston, R. J. (1968a) A simplified method of measuring membrane resistance in *Nitella translucens*. *Biochim. biophys. Acta* **150**, 518.

(1968b) The temperature dependence of the membrane potential and resistance in *Nitella translucens*. *Biochim. biophys. Acta* **150**, 640.

(1969) The membrane electrical parameters of *Nitella translucens*. *J. theoret. Biol.* **24**, 317.

Holm-Jensen, I., Krogh, A. & Wartiovaara, V. (1944) Some experiments on the exchange of potassium and sodium between single cells of the characeae and the bathing fluid. *Acta. bot. fenn.* **36**, 1.

Holtzer, H. & Sanger, J. W. (1972) Cytochalasin-B: microfilaments, cell movement and what else? *Devl. Biol.* **27**, 444.

Hope, A. B. (1951) Membrane potential differences in bean roots. *Aust. J. scient. Res. B* **4**, 265.

(1953) Salt uptake by root tissue cytoplasm: the relation between uptake and external concentration. *Aust. J. biol. Sci.* **6**, 396.

(1961) Ionic relations of cells of *Chara australis*. V. The action potential. *Aust. J. biol. Sci.* **14**, 312.

(1965) Ionic relations of cells of *Chara australis*. X. Effects of bicarbonate ions on electrical properties. *Aust. J. biol. Sci.* **18**, 789.

(1971) *Ion transport and membranes*, Butterworths, London.

Hope, A. B. & Aschberger, P. A. (1970) Effects of temperature on membrane permeability to ions. *Aust. J. biol. Sci.* **23**, 1047.

Hope, A. B. & Findlay, G. P. (1964) The action potential in *Chara*. *Pl. Cell Physiol.* **5**, 377.

Hope, A. B. & Robertson, R. N. (1953) Bioelectric experiments and the properties of plant protoplasm. *Aust. J. Sci.* **15**, 197.

Hope, A. B., Simpson, A. & Walker, N. A. (1966) The efflux of chloride from cells of *Nitella* and *Chara*. *Aust. J. biol. Sci.* **19**, 355.

Hope, A. B. & Walker, N. A. (1960) Ionic relations of cells of *Chara australis*. III. Vacuolar fluxes of sodium. *Aust. J. biol. Sci.* **13**, 277.

(1961) Ionic relations of cells of *Chara australis* R. Br. IV. Membrane potential differences and resistances. *Aust. J. biol. Sci.* **14**, 26.

Hörmann, G. (1898) *Studien über die Protoplasmaströmung bei den Characeen*, Fischer, Jena.

Inoue, I., Ueda, T. & Kobatake, Y. (1973) Structure of excitable membranes formed on the surface of protoplasmic drops isolated from *Nitella*. I. Conformation of surface membrane determined from the refractive index and from enzyme actions. *Biochim. biophys. Acta* **298**, 653.

Irwin, M. (1923) The penetration of dyes as influenced by hydrogen ion concentration. *J. gen. Physiol.* **5**, 727.

(1926) Mechanism of the accumulation of dye in *Nitella* on the basis of the entrance of dye as undissociated molecules. *J. gen. Physiol.* **9**, 561.

Jarosch, R. (1956) Plasmaströmung und chloroplastenrotation bei *Characeen. Phyton* **6**, 87.

Joliot, P. (1967) Oxygen evolution in algae illuminated with modulated light. *Brookhaven Symp. Biol.* **19**, 418.

Jost, L. (1927) Elektrische Potential differenzen an der Einzelzelle. *Sitzungsber. Heidelberger Akad. d. Wiss. (Math.-naturwiss). kl. Jahrg.*, 1927, 13 Abh.

Kamitsubo, E. (1966) Motile protoplasmic fibrils in cells of *Characeae*. I. Movement of fibrillar loops. II. Linear fibrillar structure and its bearing on protoplasmic streaming. *Proc. Japan Acad.* **42**, 507, 640.

Kamiya, N. (1962) Protoplasmic streaming. In *Encyclopedia of plant physiology*, vol. 17, part 2, 979, Springer, Berlin.

Kamiya, N. & Kuroda, K. (1956a) Artificial modification of the osmotic pressure of the plant cell. *Protoplasma.* **46**, 423.

(1956b) Velocity distribution of the protoplasmic streaming in *Nitella* cells. *Bot. Mag., Tokyo* **69**, 544.

(1957) Cell operation in *Nitella*. II. Behaviour of isolated endoplasm. *Proc. Japan Acad.* **33**, 201.

(1958a) Studies on the velocity distribution of the protoplasmic streaming in the myxomycete plasmodium. *Protoplasma* **49**, 1.

(1958b) Measurement of the motive force of the protoplasmic rotation in *Nitella. Protoplasma* **50**, 144.

(1964) Mechanical impact as a means of attacking structural organization in living cells. *Ann. Rep. sci. Wks., Fac. Sci. Osaka Univ.* **12**, 83.

(1965) Rotational protoplasmic streaming in *Nitella* and some physical properties of the endoplasm. *Proc. IV Int. Congr. Rheol.* (Part 4, Symp. *Biorheology*), p. 157, J. Wiley & Sons, New York.

(1966) Some observations of protoplasmic streaming in *Acetabularia. Bot. Mag. Tokyo* **79**, 706.

Kamiya, N. & Tazawa, M. (1956) Studies on water permeability of a single plant cell by means of transcellular osmosis. *Protoplasma* **46**, 394.

Kavanau, J. L. (1963) Protoplasmic streaming as a process of jet propulsion. *Devl Biol.* **7**, 22.

Keck, K. (1964) Culturing and experimental manipulation of *Acetabularia*. In *Methods in cell physiology*. vol. I, ed. D. M. Prescott, p. 189, Academic Press, London & New York.

Kedem, O. & Essig, A. (1965) Isotope flows and flux ratios in biological membranes. *J. gen. Physiol.* **48**, 1047.

Kedem, O. & Katchalsky, A. (1961) A physical interpretation of the phenomenological coefficients of membrane permeability. *J. gen. Physiol.* **45**, 143.

Keynes, R. D. (1969) From frog skin to sheep rumen: a survey of transport of salts and water across multicellular structures. *Q. Rev. Biophys.* **2**, 177.

Kishimoto, U. (1961) Current voltage relations in *Nitella. Biol. Bull.* **121**, 370.

(1964) Current voltage relations in *Nitella. Jap. J. Physiol.* **14**, 515.

(1966) Hyperpolarising response in *Nitella* internodes. *Pl. Cell Physiol.* **7**, 429.

Kishimoto, U. & Akabori, H. (1959) Protoplasmic streaming of an internodal cell of *Nitella flexilis*. *J. gen. Physiol.* **42**, 1167.

Kishimoto, U. & Tazawa, M. (1965a) Ionic composition of the cytoplasm of *Nitella flexilis*. *Pl. Cell Physiol.* **6**, 507.

(1965b) Ionic composition and electric response of *Lamprothamnium succinctum*. *Pl. Cell Physiol.* **6**, 529.

Kitasato, H. (1968) The influence of H^+ on the membrane potential and ion fluxes of *Nitella*. *J. gen. Physiol.* **52**, 60.

Lambers, M. H. R. (1925) The influence of temperature on protoplasmic streaming of *Characeae*. *Proc. k. ned. Akad. Wet., Amsterdam* **28**, No. 3.

Lannoye, R. J., Tarr, S. E. & Dainty, J. (1970) The effects of pH on the ionic and electrical properties of the internodal cells of *Chara australis*. *J. exp. Bot.* **21**, 543.

Lark-Horovitz, K. (1929) A permeability test with radioactive indicators. *Nature, Lond.* **123**, 277.

Larkum, A. W. D. (1968) Ionic relations of chloroplasts *in vivo*. *Nature, Lond.* **218**, 447.

Lea, E. J. A. (1963) Permeation through long narrow pores. *J. theoret. Biol.* **5**, 102.

Lefebvre, J. & Gillet, C. (1971) Effets des cations externes sur l'activité des chlorures cytoplasmic dosés par l'electrode Ag–AgCl introduite dans la cellule de *Nitella*. *Biochim. biophys. Acta* **249**, 556.

Lilley, R. M. & Hope, A. B. (1971a) Adenine nucleotide levels in cells of the marine alga, *Griffithsia*. *Aust. J. biol. Sci.* **24**, 1351.

(1971b) Chloride transport and photosynthesis in cells of *Griffithsia*. *Biochim. biophys. Acta* **226**, 161.

Ling, G. N. (1965) The physical state of water in living cell and model systems. In Forms of water in biologic systems. *Ann. N.Y. Acad. Sci.* **125**, 401.

Lucas, W. J. & Smith, F. A. (1973) The formation of alkaline and acid regions at the surface of *Chara corallina* cells. *J. exp. Bot.* **24**, 1.

Lüttge, U. (1973) Proton and chloride uptake in relation to the development of photosynthetic capacity in greening etiolated barley leaves. In *Ion transport in plants*, ed. W. P. Anderson, p. 205, Academic Press, London & New York.

Mackie, J. S. & Meares, P. (1955) The diffusion of electrolytes in a cation-exchange resin membrane. I. Theoretical. *Proc. R. Soc. A.* **232**, 498.

MacRobbie, E. A. C. (1962) Ionic relations of *Nitella translucens*. *J. gen. Physiol.* **45**, 861.

(1964) Factors affecting the fluxes of potassium and chloride ions in *Nitella translucens*. *J. gen. Physiol.* **47**, 859.

(1966) Metabolic effects on ion fluxes in *Nitella translucens*. I. Active influxes. *Aust. J. biol. Sci.* **19**, 363.

(1969) Ion fluxes to the vacuole of *Nitella translucens*. *J. exp. Bot.* **20**, 236.

(1970a) Quantized fluxes of chloride in the vacuole of *Nitella translucens*. *J. exp. Bot.* **21**, 335.

(1970b) The active transport of ions in plant cells. *Q. Rev. Biophys.* **3**, 251.

(1971) Vacuolar fluxes of chloride and bromide in *Nitella translucens*. *J. exp. Bot.* **22**, 487.

(1973) Vacuolar ion transport in *Nitella*. In *Ion Transport in Plants*, ed. W. P. Anderson, p. 431, Academic Press, London and New York.

MacRobbie, E. A. C. & Dainty, J. (1958) Ion transport in *Nitellopsis obtusa*. *J. gen. Physiol.* **42**, 335.

MacRobbie, E. A. C. & Fensom, D. S. (1969) Measurements of electro-osmosis in *Nitella translucens*. *J. exp. Bot.* **20**, 466.

Mailman, D. S. & Mullins, L. J. (1966) The electrical measurement of chloride fluxes in *Nitella*. *Aust. J. biol. Sci.* **19**, 385.

Marchant, H. J. & Pickett-Heaps, J. D. (1970) Ultrastructure and differentiation of *Hydrodictyon reticulatum*. I. Mitosis in the coenobium. *Aust. J. biol. Sci.* **23**, 1173.

(1972) Ultrastructure and differentiation of *Hydrodictyon reticulatum*. VI. Formation of the germ net. *Aust. J. biol. Sci.* **25**, 1199.

Meares, P. & Ussing, H. H. (1959) The fluxes of sodium and chloride ions across a cation-exchange resin membrane. Part 1. Effect of a concentration gradient. Part 2. Diffusion with electric current. *Trans. Faraday Soc.* **55**, 142 and 244.

Mercer, M. J. & Mercer, F. V. (1971) Studies on the comparative physiology of *Chara corallina*. III. Nitrogen relations of internodal cell components during internodal cell expansion. *Aust. J. Bot.* **19**, 1.

Metlicka, R. & Rybova, R. (1967) Oscillations of the trans-membrane potential difference in the alga *Hydrodictyon reticulatum*. *Biochim. biophys. Acta* **135**, 563.

Meyer, A. (1891) Notiz über die Zusammensetzung des Zellsafter von *Valonia utricularis*. *Ber. deutsch. bot. Ges.* **9**, 77.

Mitchell, P. (1966) Chemiosmotic coupling in oxidative and photosynthetic phosphorylation. *Biol. Rev.* **41**, 445.

(1967) Translocations through natural membranes. *Adv. Enzymol.* **29**, 33.

Mullins, L. J. (1962) Efflux of chloride ions during the action potential of *Nitella*. *Nature, London* **196**, 986.

Mullins, L. J. & Brooks, S. C. (1939) Radioactive ion exchanges in living protoplasm. *Science, Washington* **90**, 256.

Nagai, R. & Kishimoto, U. (1964) Cell wall potential in *Nitella*. *Pl. Cell Physiol.* **5**, 21.

Nagai, R. & Rebhun, L. I. (1966) Cytoplasmic microfibrils in streaming *Nitella* cells. *J. ultrastruct. Res.* **14**, 571.

Nagai, R. & Tazawa, M. (1962) Changes in resting potential and ion absorption induced by light in a single plant cell. *Pl. Cell Physiol.* **3**, 323.

Nägeli, C. (1860) *Beiträge zur wiss. Bot.* **2**, 62.

Nägeli, C. & Cramer, C. (1855) *Pflanzenphysiologische Untersuchungen*, vol. I, Schulthess, Zurich.

Nathansohn, A. (1903) Ueber Regulationserscheinungen im Stoffaustausch *Jahrb. Wiss. Bot.* **38**, 249.

Nichols, S. P. (1925) The effect of wounds upon the rotation of the protoplasm in the internodes of *Nitella*. *Bull. Torrey bot. Club* **52**, 351.

Nishizaki, Y. (1963) Bioelectric potential of *Chara* under intermittent illumination. *Pl. Cell Physiol.* **4**, 353.

(1968) Light induced changes of bioelectric potential in *Chara*. *Pl. Cell Physiol.* **9**, 377.

Northcote, D. H., Goulding, K. J. & Horne, R. W. (1960) Chemical composition and structure of the cell wall of *Hydrodictyon africanum*. *Biochem. J.* **77**, 503.

Ockam, Guillermus (1487) Quotlibeta Septem. Impressa Parisii Magistri Petri Rubei. *Brunet* **4**, 154.

Oda, K. (1961) The electrical constant in *Chara braunii*. *Sci. rep. Tohoku Univ., ser. IV, Biol.* **27**, 187.

(1962) Polarised and depolarised states of the membrane in *Chara braunii*, with special reference to the transition between the two states. *Sci. rep. Tohoku Univ., ser. IV, Biol.* **28**, 1.

Osterhout, W. J. V. (1914) Uber der Temperaturkoeffizienten des elektrischen Leitvermögens im lebenden und toten Gewebe. *Biochem. Z.* **67**, 272.

(1918) Comparative studies on respiration. I. Introduction. *J. gen. Physiol.* **1**, 171.

(1920) The mechanism of injury and recovery. *J. gen. Physiol.* **3**, 15.

(1922) Direct and indirect determinations of permeability. *J. gen Physiol.* **4**, 275.

(1931) Physiological studies of single plant cells. *Biol. Rev.* **6**, 369.

(1949) Movements of water in cells of *Nitella* and Transport of water from concentrated to dilute solutions in cells of *Nitella*. *J. gen. Physiol.* **32**, 553 and 559.

Osterhout, W. J. V., Damon, E. B. & Jacques, A. G. (1927) Dissimilarity of inner and outer protoplasmic surfaces in *Valonia*. *J. gen. Physiol.* **11**, 193.

Osterhout, W. J. V. & Harris, E. S. (1928) Protoplasmic asymmetry in *Nitella* as shown by bioelectric measurements. *J. gen. Physiol.* **11**, 391.

(1929) The concentration effect in *Nitella*. *J. gen. Physiol.* **12**, 761.

Overbeek, J. Th. (1956) The donnan equilibrium. *Prog. Biophys. biophys. Chem.* **6**, 57.

Overton, E. (1899) Uber die allgemeinen Eigenschaften der Zelle, ihre vermutlichen Ursachen und ihre Bedeutung fur der Physiologie. *Vjschr. Naturforsch. Ges. Zurich* **44**, 88.

(1900) Studien über die Aufnahme der Anilinfarben durch die lebende Zelle. *Jahrb. f. wiss. Bot.* **34**, 669.

(1902) Beiträge zur allgemeinen Muskel-u-Nervenphysiologie. *Pflugers Arch. ges. Physiol.* **92**, 115.

Paganelli, C. V. & Solomon, A. K. (1957) The rate of exchange of tritiated water across the human red cell membrane. *J. gen. Physiol.* **41**, 259.

Pallaghy, C. K. & Scott, B. I. H. (1969) The electrochemical state of cells of broad bean roots. II. Potassium kinetics in excised root tissue. *Aust. J. biol. Sci.* **22**, 585.

Paszewski, A., Stolarek, J. & Gebal, T. (1968) Ionic relations and electrophysiology of single cells of Characeae. Part I. Investigations on electric potentials and resistance in cells of *Nitellopsis obtusa*. *Acta Soc. bot. Pol.* **37**, 327.

Pfeffer, W. (1877) *Osmotische Untersuchungen*, Leipzig.

Pickett-Heaps, J. D. (1967) Ultrastructure and differentiation in *Chara* sp. I. Vegetative cells. *Aust. J. biol. Sci.* **20**, 539.

Pidot, A. L. & Diamond, J. M. (1964) Streaming potentials in a biological membrane. *Nature, London* **201**, 701.

Plowe, J. Q. (1931) Membranes in the plant cell. I. Morphological membranes at protoplasmic surfaces. II. Localisation of differential permeability in the plant protoplast. *Protoplasma* **12**, 196 and 221.

Poglazov, B. F. (1966) Chapter 8 in *Structure and functions of contractile proteins*, p. 231, Academic Press, New York.

Polya, G. M. (1968) Nucleotide metabolism and ion transport in plant cells. Ph. D. Thesis, Flinders University, South Australia.

Prescott, D. M. & Zeuthen, E. (1953) Comparison of water diffusion and water filtration across cell surfaces. *Acta physiol. Scand.* **28**, 77.

Pressman, B. C., Harris, E. J., Jagger, W. S. & Johnson, J. H. (1967) Antibiotic-mediated transport of alkali ions across lipid barriers. *Proc. nat. Acad. Sci.* **58**, 1949.

Proctor, V. W. (1971) *Chara globularis thuillier* (=*C. fragilis Desvaux*): Breeding patterns within a cosmopolitan complex. *Limnol. and Oceanog.* **16**, 422.

Raven, J. A. (1967) Ion transport in *Hydrodictyon africanum*. *J. gen. Physiol.* **50**, 1607.

(1968a) The mechanism of photosynthetic use of bicarbonate by *Hydrodictyon africanum*. *J. exp. Bot.* **19**, 193.

(1968b) The linkage of light-stimulated Cl influx to K and Na influxes in *Hydrodictyon africanum*. *J. exp. Bot.* **19**, 233.

(1968c) The action of phlorizin on photosynthesis and light stimulated ion transport in *Hydrodictyon africanum*. *J. exp. Bot.* **19**, 712.

(1969a) Action spectra for photosynthesis and light-stimulated ion transport processes in *Hydrodictyon africanum*. *New Phyt.* **68**, 45.

(1969b) Effects of inhibitors on photosynthesis and the active influxes of K and Cl in *Hydrodictyon africanum*. *New Phyt.* **68**, 1089.

(1970a) The role of cyclic and pseudo-cyclic photophosphorylation in photosynthetic $^{14}CO_2$ fixation in *Hydrodictyon africanum*. *J. exp. Bot.* **21**, 1.

(1970b) Exogenous inorganic carbon sources in plant photosynthesis. *Biol. Rev.* **45**, 167.

(1970c) Ouabain-insensitive K-influx in *Hydrodictyon africanum*. *Planta, Berlin* **97**, 28.

(1971) Inhibitor effects on photosynthesis, respiration and ion transport in *Hydrodictyon africanum*. *J. memb. Biol.* **6**, 89.

Raven, J. A., MacRobbie, E. A. C. & Neuman, J. (1969) The effect of Dio-9 on photosynthesis and ion transport in *Nitella, Tolypella* and *Hydrodictyon*. *J. exp. Bot.* **20**, 221.

Remington, R. E. (1928) The high frequency wheatstone bridge as a tool in cytological studies. *Protoplasma* **5**, 338.

Rent, R. K., Johnson, R. A. & Barr, C. E. (1972) Net H^+ influx in *Nitella clavata*. *J. memb. Biol.* **7**, 231.

Rich, G. T., Sha'afi, R. I., Barton, T. C. & Solomon, A. K. (1967) Permeability studies on red cell membranes of dog, cat and beef. *J. gen. Physiol.* **50**, 2391.

Richards, J. L. & Hope, A. B. (1974) The role of protons in determining membrane electrical characteristics in *Chara corallina. J. memb. Biol.* **16**, 121.

Robertson, R. N. (1968) *Protons, electrons, phosphorylation and active transport,* Cambridge University Press, Cambridge, England.

Robertson, R. N. & Turner, J. S. (1945) Studies on the metabolism of plant cells IV. The effects of cyanide on the accumulation of potassium chloride and on respiration; the nature of salt respiration. *Aust. J. biol. med. Sci.* **23**, 63.

Robinson, J. B. (1969a) Sulphate influx in characean cells. I. General characteristics. *J. exp. Bot.* **20**, 201.

(1969b) Sulphate influx in characean cells. II. Links with light and metabolism in *Chara australis. J. exp. Bot.* **20**, 212.

Romijn, G. (1931) Uber den Einflus der Temperatur auf die Protoplasmaströmung bei *Nitella flexilis. Proc. k. ned. Akad. Wet.,* Amsterdam **34**, 163.

Round, F. E. (1971) The taxonomy of the chlorophyta. II. *Br. phycol. J.* **6**, 235.

Saddler, H. W. D. (1970) The membrane potential of *Acetabularia mediterranea. J. gen. Physiol.* **55**, 802.

Sandan, T. & Somura, T. (1959) Effect of A.T.P. on the rate of the protoplasmic streaming in *Nitella. Bot. Mag., Tokyo* **72**, 337.

Schilde, C. (1966) Zur Wirkung des Lichtes auf des Ruhepotential der grunen Pflanzenzelle. *Planta, Berlin* **71**, 184.

Sha'afi, R. I., Rich, G. T., Sidel, V. W., Bossert, W. & Solomon, A. K. (1967) The effect of the unstirred layer on human red cell water permeability. *J. gen. Physiol.* **50**, 1377.

Sibaoka, T. (1966) Action potentials in plant cells. In *Nervous and Hormonal Mechanisms of Integration, Symp. Soc. exp. Biol.* **20**, 49.

Sidel, V. W. & Solomon, A. K. (1957) Entrance of water into human red cells under an osmotic pressure gradient. *J. gen. Physiol.* **41**, 243.

Skierczynska, J. (1968) Some of the electrical characteristics of the cell membrane of *Chara australis. J. exp. Bot.* **19**, 389.

Skou, J. C. (1965) Enzymatic basis for active transport of Na^+ and K^+ across cell membrane. *Physiol. Rev.* **45**, 596.

Slack, H. (1834) Exposition des tissus élémentaires des plantes, avec quelques exemples de circulation végétales. *Annls. sci. Nat.* **1**, Abt II, 193 and 271.

Smith, F. A. (1966) Active phosphate uptake by *Nitella translucens. Biochim. biophys. Acta* **126**, 94.

(1967a) Links between glucose uptake and metabolism in *Nitella translucens. J. exp. Bot.* **18**, 348.

(1967b) Rates of photosynthesis in characean cells. 1. Photosynthetic $^{14}CO_2$ fixation by *Nitella translucens. J. exp. Bot.* **18**, 509.

(1968a) Metabolic effects on ion fluxes in *Tolypella intricata. J. exp. Bot.* **19**, 422.

(1968b) Rates of photosynthesis in characean cells. II. Photosynthetic $^{14}CO_2$ fixation and ^{14}C-bicarbonate uptake by characean cells. *J. exp. Bot.* **19**, 207.

(1970) The mechanism of chloride transport of characean cells. *New Phytol.* **69**, 903.

(1971) The control of chloride transport in *Chara* by pH gradient. *Proc. 1st Eur. Biophys. Congress*, eds E. Broda, A. Locker & H. Springer-Lederer, p. 429, Verlag der Weiner Medizinschen Akademie, Vienna.

Smith, F. A. & West, K. R. (1969) A comparison of the effects of metabolic inhibitors on chloride uptake and photosynthesis in *Chara corallina*. *Aust. J. biol. Sci.* **22**, 351.

Solomon, A. K. (1968) Characterization of biological membranes by equivalent pores. *J. gen. Physiol.* **51**, 335S.

Spanner, D. C. (1964) *Introduction to thermodynamics*, Academic Press, London.

Spanswick, R. M. (1970a) The effects of bicarbonate ions and external pH on the membrane potential and resistance of *Nitella translucens*. *J. memb. Biol.* **2**, 59.

Spanswick, R. M. (1970b) Electrophysiological techniques and the magnitudes of the membrane potentials and resistances of *Nitella translucens*. *J. exp. Bot.* **21**, 617.

(1973) Evidence for an electrogenic ion pump in *Nitella translucens*. I. The effects of pH, K^+, Na^+, light and temperature on the membrane potential and resistance. *Biochim. biophys. Acta* **288**, 73.

Spanswick, R. M. & Costerton, J. W. F. (1967) Plasmodesmata in *Nitella translucens*: structure and electrical resistance. *J. cell Sci.* **2**, 451.

Spanswick, R. M., Stolarek, J. & Williams, E. J. (1967) The membrane potential of *Nitella translucens*. *J. exp. Bot.* **18**, 1.

Spanswick, R. M. & Williams, E. J. (1964) Electrical potentials and the Na, K and Cl concentrations in the vacuole and cytoplasm of *Nitella translucens*. *J. exp. Bot.* **15**, 193.

(1965) Ca fluxes and membrane potentials in *Nitella translucens J. exp. Bot.* **16**, 463.

Spear, D. G., Barr, J. K. & Barr, C. E. (1969) Localization of hydrogen ion and chloride fluxes in *Nitella*. *J. gen. Physiol.* **54**, 397.

Spiegler, K. S. (1958) Transport processes in ionic membranes. *Trans. Faraday Soc.* **54**, 1408.

Spyropoulos, C. S. (1972) Some observations of the electrical properties of biological membranes. In *Membranes*, vol. I, *Macroscopic systems and models*, ed. G. Eisenman, p. 267, Marcel Dekker Inc., New York.

Stein, W. D. (1967) *The movement of molecules across cell membranes*, Academic Press, New York and London.

Steudle, E. & Zimmermann, U. (1971a) Zellturgor und selektiver Ionen transport bei *Chaetomorpha linum*. *Z. Naturf.* **26b**, 1276.

(1971b) Hydrolische Leitfähigkeit von *Valonia utricularis*. *Z. Naturf.* **26b**, 1302.

Steward, F. C. & Sutcliffe, J. F. (1959) Plants in relation to inorganic salts. In *Plant Physiology* vol. II, ed. F. C. Steward, p. 253, Academic Press, New York and London.

Stewart, G. N. (1899) The behaviour of the haemoglobin and electrolytes of the coloured corpuscles when blood is laked. *J. Physiol.* **24**, 211.

Strunk, T. H. (1970) Vacuolar perfusion technique for *Nitella* internodal cells. *Science*, Washington **169**, 84.

Takata, M. (1961) Studies on the protoplasmic streaming in the marine alga *Acetabularia*. *Ann. Rep. Sci. Works, Fac. Sci., Osaka Univ.* **9**, 63.

Tay, D. K. C. & Findlay, G. P. (1972) Permeability of the duodenum of the toad to non-electrolytes. *Aust. J. biol. Sci.* **25**, 931.

Taylor, C. V. & Whitaker, D. M. (1926) A measurable potential difference between the cell interior and the outside medium. *Carnegie Inst. Year Book* **25**, 248.

— (1928) Potentiometric determinations in the protoplasm and cell sap of *Nitella*. *Protoplasma* **3**, 1.

Tazawa, M. (1957) Neue Methode zur Messung des osmotischen Wertes einer Zelle. *Protoplasma* **48**, 342.

— (1964) Studies on *Nitella* having artificial cell sap. I. Replacement of the cell sap with artificial solutions. *Pl. Cell Physiol.* **5**, 33.

Tazawa, M. & Kamiya, N. (1966) Water permeability of a characean internodal cell with special reference to its polarity. *Aust. J. biol. Sci.* **19**, 399.

Tazawa, M. & Kishimoto, U. (1968) Cessation of cytoplasmic streaming of *Chara* internodes during action potential. *Pl. Cell Physiol.* **9**, 361.

Tazawa, M. & Nagai, R. (1966) Studies on osmoregulation of *Nitella* internode with modified cell saps. *Z. Pflantzen.* **54**, 333.

Teorell, T. (1949) Membrane electrophoresis in relation to bioelectrical polarization effects. *Arch. Sci. Physiol.* **3**, 205.

— (1959) Electrokinetic membrane processes in relation to properties of excitable tissues I and II. *J. gen. Physiol.* **42**, 831 and 847.

Thain, J. F. (1967) *Principles of Osmotic Phenomena*, R. Inst. Chem., London.

— (1973) The flux-ratio for potassium in *Chara corallina*. In *Ion transport in plants*, ed. W. P. Anderson, p. 77, Academic Press, London and New York.

Thomas, D. A. (1971) The regulation of stomatal aperture in tobacco leaf epidermal strips. III. The effect of ATP. *Aust. J. biol. Sci.* **24**, 689.

Tyree, M. T. (1968) Determination of transport constants of isolated *Nitella* cell walls. *Can. J. Bot.* **46**, 317.

Tyree, M. T. & Spanner, D. C. (1969) A reappraisal of thermodynamic transport coefficients in *Nitella* cell walls. *Can. J. Bot.* **47**, 1497.

Umrath, K. (1930) Untersuchungen über Plasma und Plasmaströmung an Characeen. IV. Potentialmessungen an *Nitella mucronata* mit besonderer Berücksichtigung der Erregungserscheinung. *Protoplasma* **9**, 576.

— (1932) Die Bildung von Plasmalemma (Plasmahaut) bei *Nitella mucronata*. *Protoplasma* **16**, 173.

— (1934) Der Einfluss der Temperatur auf das elektrische Potential, den Aktionsström und die Protoplasmaströmung bei *Nitella mucronata*. *Protoplasma* **21**, 329.

— (1938) Über das elektrische Potential und Aktionströme von *Valonia macrophysa*. *Protoplasma* **31**, 184.

Ussing, H. H. (1949) The distinction by means of tracers between active transport and diffusion. *Acta. physiol. Scand.* **19**, 43.

Ussing, H. H. & Zerahn, K. (1951) Active transport of sodium as the source of the electric current in the short-circuited isolated frog skin. *Acta. physiol. Scand.* **23**, 111.

Velten, W. (1876) Die Enwirkung der Temperatur auf die Protoplasmabewung. *Flora* 59, 209.

Volkov, G. (1964) On the change in resting potential of individual *Nitella flexilis* cells induced by light. *Akad. Nauk. S.S.S.R. Doklad.* 155, 1224.

Vorobiev, L. N. (1967) Potassium ion activity in the cytoplasm and the vacuole of cells of *Chara* and *Griffithsia*. *Nature, London* 216, 1325.

Vorobiev, L. N., Koltunov, Yu. B., Kurella, G. A. & Li, Su-Yuen (1965) Mean activity of potassium salts in cell sap of *Nitella mucronata in situ*. *Biophysics* 10, 591.

Vorobiev, L. N., Radenovich, Ch. N., Khitrov, Yu. A. & Yaglova, L. G. (1967) Investigation of jumps in biopotentials on insertion of a microelectrode into the vacuole of the cells of *Nitella mucronata*. *Biophysics* 12, 1163.

Vorobieva, I. A. & Vorobiev, L. N. (1965) Action of ATP on resting potential and movement of protoplasm of *Nitella mucronata*. *Biofizika* 10, 1007.

Vredenberg, W. J. (1969) Light induced changes in membrane potential of algal cells associated with photosynthetic electron transport. *Biochim. biophys. Res. Comm.* 37, 785.

(1970a) Application of the voltage-clamp technique for measuring the quantum efficiency of light-induced potential changes in *Nitella translucens*. *Biochim. biophys. Acta* 216, 431.

(1970b) Chlorophyll *a* fluorescence induction and changes in the electrical potential of the cellular membranes of green plant cells. *Biochim. biophys. Acta* 223, 230.

(1973) Energy control of ion fluxes in *Nitella* as measured by changes in potential, resistance and current-voltage characteristics of the membranes. In *Ion Transport in Plants*, ed. W. P. Anderson, p. 153, Academic Press, London and New York.

de Vries, H. (1884) Eine Methode zur Analyse der Turgorkraft. *Jahrb. f. wiss. Bot.* 14, 427.

Walker, D. A. & Crofts, A. R. (1970) Photosynthesis. *Ann. Rev. Biochem.* 39, 389.

Walker, N. A. (1955) Microelectrode experiments on *Nitella*. *Aust. J. biol. Sci.* 8, 476.

(1960) The electric resistance of the cell membranes in a *Chara* and a *Nitella* species. *Aust. J. biol. Sci.* 13, 468.

(1973) Discussion on MacRobbie's paper in *Ion Transport in Plants*, ed. W. P. Anderson, p. 459, Academic Press, London.

Walker, N. A. & Bostrom, T. E. (1973) Intercellular movement of chloride in *Chara* – a test of models for chloride influx. In *Ion transport in plants*, ed. W. P. Anderson, p. 447, Academic Press, London.

Walker, N. A. & Hope, A. B. (1969) Membrane fluxes and electrical conductance in characean cells. *Aust. J. biol. Sci.* 22, 1179.

Wartiovaara, V. (1942) Über die Temperaturabhangigkeit der Protoplasmapermeabilität. *Annales Bot. Soc. Zool. -Bot. Fenn. Vanamo* T16, 1.

(1949) The permeability of the plasma membranes of *Nitella* to normal primary alcohols at low and intermediate temperatures. *Physiol. Plant* 2, 184.

Wartiovaara, V. & Collander, R. (1960) Permeabilitätstheorien. *Proto-plasmatologia*, 2C8d, 1.

Wessels, N. K., Spooner, B. S., Ash, J. F., Bradley, M. D., Luduena, M. A., Taylor, E. L., Wrenn, J. T. & Yamada, K. M. (1971) Microfilaments in cellular and developmental processes. *Science*, Washington 171, 135.

Wiebelhaus, V. D., Sung, C. P., Helander, H. F., Shah, G., Blum, A. L. & Sachs, G. (1971) Solubilization of anion ATPase from *Necturus* oxynctic cells. *Biochim. biophys. Acta* 241, 49.

Williams, E. J., Johnston, R. J. & Dainty, J. (1964) The electrical resistance and capacitance of the membranes of *Nitella translucens*. *J. exp. Bot.* 15, 1.

Williamson, R. E. (1972). A light-microscope study of the action of cyto-chalasin B on the cells and isolated cytoplasm of the Characeae, *J. cell. Sci.* 10, 811.

Wodehouse, R. P. (1917) Direct determinations of permeability. *J. biol. Chem.* 29, 453.

Wood, R. D. (1952) The characeae, 1951. *Bot. Rev.* 18, 317.

Wood, R. D. & Imahori, K. (1965) *A revision of the Characeae*, Verlag Von J. Cramer, Weinheim.

Wright, E. M. & Diamond, J. M. (1969) Patterns of non-electrolyte permeability. *Proc. R. Soc. B.* 172, 227.

Young, J. Z. (1936) Structure of nerve fibres and synapses in some inverte-brates. *Cold Spr. Harb. Symp. quant. Biol.* 4, 1.

Zimmermann, U. & Steudle, E. (1970) Bestimmung von Reflexions coeffizienten an der Membran der Alge *Valonia utricularis*. *Z. Naturf.* 25b, 500.

  (1971) Effects of potassium concentration and osmotic pressure of sea water on the cell-turgor pressure of *Chaetomorpha linum*. *Marine Biol.* 11, 132.

Zwolinski, B. J., Eyring, H. & Reese, C. E. (1949) Diffusion and membrane permeability, I. *J. phys. colloid Chem.* 53, 1426.

Index